Lecture Notes in Statistics

Volume 1028

For further volumes:
http://www.springer.com/series/694

Lecture N

Volume 1

Edited by

P. Bickel
P. Diggle
S. E. Fienberg
U. Gather
I. Olkin
S. Zeger

For further volumes:

B. K. Sinha · N. K. Mandal
Manisha Pal · P. Das

Optimal Mixture Experiments

Springer

B. K. Sinha
Indian Statistical Institute
Kolkata
West Bengal
India

Manisha Pal
Department of Statistics
University of Calcutta
Kolkata
India

N. K. Mandal
Department of Statistics
University of Calcutta
Kolkata
India

P. Das
Department of Statistics
University of Kalyani
Kalyani, Nadia
West Bengal
India

ISSN 0930-0325 ISSN 2197-7186 (electronic)
ISBN 978-81-322-1785-5 ISBN 978-81-322-1786-2 (eBook)
DOI 10.1007/978-81-322-1786-2
Springer New Delhi Heidelberg New York Dordrecht London

Library of Congress Control Number: 2014936178

Printed on acid-free paper

Springer is part of Springer Science+Business Media (www.springer.com)

We dedicate this monograph to all the researchers in this fascinating topic whose valuable contributions have paved the way of our thought process in writing this monograph. All of them have been with us in spirit and action. We fondly hope we have not disappointed them

Foreword

It gives me much pleasure to welcome this comprehensive work on the optimality aspects of Mixture Designs. Kiefer formulated various optimality criteria and laid foundations for the study of optimal designs—both exact and approximate. Scheffé initiated the study of mixture designs and introduced a model for these designs. The current monograph brings these together in a cohesive way. It deals with Scheffé's model and some other models. Optimality aspects studied include: (a) optimal designs for the estimation of parameters in mixture models and (b) optimal designs for optimal mixtures under mixture models. This is followed by applications of mixture experiments in various fields such as agriculture and pharmaceuticals. The monograph concludes with a study of several variants and extensions. It also gives directions for further research in this area of study which has just opened up. The authors have taken a significant step in promoting research in this area of study by putting together all known work in this area and by indicating directions for further work. They deserve our gratitude for this important contribution.

Waterloo, Canada, February 2014 Kirti R. Shah

Preface

Two of us had necessary research experience in the broad area of optimal designs. One of us had necessary exposition in the area of response surface-related optimal designs. The remaining one had adequate experience in handling factorial designs. Two of us ventured into this emerging area of optimal mixture designs several years back. The other two joined hands and strengthened the research collaboration to the extent that all of us together could see the emergence of a research-level monograph within a reasonable time frame.

We have tried to present and explain, in our own way, the techniques needed for handling the optimal designing problems related to the general area of mixture models and specific areas of applications of such models.

We will consider our efforts rewarded if the readers find the presentations in order and derive enough creative interest in pursuing the research topics further.

Professor Kirti R. Shah has showered unbounded research opportunity on two of us for over 20 long years or so. He has obliged us by very kindly agreeing to write the Foreword for this monograph.

We have a very special point to make. The whole exercise was academically challenging to one of us who had no exposure to this area of research. But with sheer interest, enthusiasm, and dedication, this special collaborator exceeded all our expectations and rightfully deserved a position on the front page of the publication.

Kolkata, February 2014

B. K. Sinha
N. K. Mandal
Manisha Pal
P. Das

Acknowledgments

This monograph is the end product of a research project entitled *Optimum Regression Designs* that was undertaken by us in the Department of Statistics, University of Calcutta, and was supported by the UPE (University with Potential for Excellence) Grant received from the University Grants Commission, India. We are thankful to the department and our colleagues for providing a congenial and supportive atmosphere during the preparation of the monograph. We also thank Prof. S. P. Mukherjee, our revered teacher and Centenary Professor of Statistics in the department [now retired], for his continued interest and encouragement.

At different stages, three workshops on this theme were conducted by us at (i) Kalyani University's Department of Statistics, (ii) Indian Agricultural Statistics Research Institute, New Delhi, and (iii) Calcutta University's Department of Statistics. We thank the participants for their interest and questions/queries which helped us shape the presentation material for this monograph.

We feel highly indebted to Mr. Aparesh Chatterjee for giving time and energy to create the latex file of the entire manuscript. We have been corresponding with him over the last one year in all permutations and combinations—at times even from overseas—with frequent and quick changes in the draft. We are truly happy to note that he has favored us by not departing in the middle of the chaos! He indeed deserves a special mention and our thanks for bearing with our demands with all smiles and extreme courtesy at all stages!!

Finally, with heartfelt thanks, we express our gratitude to our family members for their love and unlimited support, and our appreciation to Ms. Sagarika Ghosh of Springer for her valued help at various stages.

Kolkata, February 2014

Bikas Kumar Sinha
Nripes Kumar Mandal
Manisha Pal
Premadhis Das

Contents

About the Authors

Professor Bikas Kumar Sinha is a retired professor of statistics at Indian Statistical Institute, Calcutta, India. He has traveled extensively within India and abroad for collaborative research and with teaching assignments. He has served as a "UN Expert on Mission" and also as a consultant for USEPA. He is a "Summer at Census" visitor at the US Census Bureau. He has more than 130 research publications in areas like linear models and optimal designs, survey sampling theory, and statistical methodologies. He is the co-author of two research monographs published under Springer-Verlag Lecture Notes in Statistics Series (1989, 2002).

Professor Nripes Kumar Mandal is a senior faculty in the Department of Statistics, University of Calcutta, India. He has been working in the area of Design of Experiments since 1977. He has visited several universities/institutes in India and abroad for teaching/collaborative research and as a resource person/invited speaker/keynote speaker at Workshops/Conferences/Symposia. He has more than 60 research articles published in peer-reviewed journals. He is the co-author of a research monograph published under Springer-Verlag Lecture Notes in Statistics Series (2002).

Professor Manisha Pal teaches at the Department of Statistics, University of Calcutta, India. She has undertaken teaching and collaborative research assignments in Indian and foreign universities. She has been a resource person/invited speaker at workshops/conferences/symposia held in India and abroad. She has more than 85 published research articles in the areas of inventory control, reliability inference, skewed distributions, mixture experiments and data analysis.

Professor Premadhis Das is a senior faculty at the Department of Statistics, University of Kalyani, India. He has been working in the area of design of experiments for more than 30 years, and has visited many universities for teaching/collaborative research and as a resource person/invited speaker at workshops/conferences/symposia. He has research publications in many national and international journals of repute.

Chapter 1
Mixture Models and Mixture Designs: Scope of the Monograph

Abstract We introduce standard mixture models and standard mixture designs as are well known in the literature (vide Cornell 2002). Some of the less known models are also introduced briefly. Next we explain the frameworks of exact and approximate [or, continuous] mixture designs. We mention about known applications of mixture experiments in agriculture, food processing, and pharmaceutical studies. We also provide a chapter-wise brief summary of the contents covered in the monograph.

Keywords Scheffé's homogeneous mixture models · Becker's mixture model · Draper–St. John's mixture model · Simplex lattice designs · Simplex centroid designs · Axial designs · Exact designs · Approximate designs · Applications · Agriculture · Pharmacy · Food processing

1.1 Introduction

This monograph features state-of-the-art research findings on various aspects of mixture experiments, mainly from the point of view of *optimality*. We have freely consulted available books and journals on mixture experiments and optimal experimental designs. There is no denying the fact that a considerable number of research articles has been published in this specific area dealing with mixture model specifications and related data analyses; however, emphasis on finding optimal mixture experiments has been relatively less pronounced. With a thorough understanding of the tools and techniques in the study of optimal designs [in discrete and continuous design settings], we ventured into this relatively new area of research a few years back and we were fascinated by the niceties of the elegant results—already known in the literature and further researched out by our team. We are happy to work on this monograph and bring it to the attention of optimal design theorists in a most comprehensive manner—covering basic and advanced results in the area of optimal mixture experiments.

B. K. Sinha et al., *Optimal Mixture Experiments*, Lecture Notes in Statistics 1028,
DOI: 10.1007/978-81-322-1786-2_1,

1.2 Mixture Models

Let $\boldsymbol{x} = (x_1, x_2, \ldots, x_q)$ denote the vector of proportions of q mixing components and $\eta(\boldsymbol{x})$ be the corresponding mean response. The factor space is a simplex, given by

$$\mathcal{X} = \{\boldsymbol{x} = (x_1, \ldots, x_q) : x_i \geq 0, \quad i = 1, 2, \ldots, q; \ \sum_{i=1}^{q} x_i = 1\}. \tag{1.2.1}$$

Scheffé (1958) introduced the following models in *canonical* forms of different degrees to represent the mean response function $\eta(\boldsymbol{x})$:

$$\text{Linear: } \eta(\boldsymbol{x}) = \sum_i \beta_i x_i \tag{1.2.2}$$

$$\text{Quadratic: } \eta(\boldsymbol{x}) = \sum_i \beta_i x_i + \sum_{i<j} \beta_{ij} x_i x_j \tag{1.2.3}$$

$$\text{Cubic: } \eta(\boldsymbol{x}) = \sum_i \beta_i x_i + \sum_{i<j} \beta_{ij} x_i x_j + \sum_{i<j<k} \beta_{ijk} x_i x_j x_k \tag{1.2.4}$$

$$\text{Special Cubic: } \eta(\boldsymbol{x}) = \sum_i \beta_i x_i + \sum_{i<j} \beta_{ij} x_i x_j + \sum_{i<j<k} \beta_{ijk} x_i x_j x_k + \sum_{i<j} \beta_{ij} x_i x_j (x_i - x_j). \tag{1.2.5}$$

In the above, we have used generic notations for the model parameters in different versions of mixture models. Using the identity $\Sigma x_i = 1$, model (1.2.3) can be converted to a canonical *homogeneous* quadratic model:

$$\eta(\boldsymbol{x}) = \sum_i \beta_{ii} x_i^2 + \sum_{i<j} \beta_{ij} x_i x_j. \tag{1.2.6}$$

In the present study, we shall be concerned with the canonical models (1.2.2) and (1.2.3) or (1.2.6) or some other equivalent versions of it. There are other types of mixture models introduced in the literature. In order to differentiate them, we may refer to the above models as 'standard mixture models'. It is to be noted that the use of the identity $\Sigma x_i = 1$ may lead to remodeling very deceptively. For example, the quadratic model (1.2.3) may be modified to the cubic model, thus inviting more parameters [with intriguing parametric relations] and more design points for estimation. Also proper interpretation of the parameters may be a bit confusing and complicated. Note that unlike in the usual regression models, the constant term β_0 has been dropped from all mixture models, as otherwise, the β-parameters become non-estimable. Further to this, in the mixture model setup, β_0 does not have any obvious interpretation like the intercept!

As was mentioned above, there are some other 'non-standard' mixture models introduced and studied in the literature. We will deal with symmetrized versions

of two such models viz. Becker's homogeneous model of degree one (1968) and Draper–St. John's model (1977).
Becker's model is given by

$$\eta(\boldsymbol{x}) = \beta_1 x_1 + \beta_2 x_2 + \cdots + \beta_q x_q + \beta_{12}\frac{x_1 x_2}{x_1 + x_2} + \cdots + \beta_{q-1q}\frac{x_{q-1}x_q}{x_{q-1} + x_q}$$
$$0 \leq x_i \leq 1, \forall i;\ x_i + x_j > 0,\ \forall i < j. \tag{1.2.7}$$

As a matter of fact, Becker introduced a general representation of (1.2.7) as is given below:

$$\eta(\boldsymbol{x}) = \sum_i \beta_i x_i + \sum_{i<j} \beta_{ij}\frac{x_i x_j}{x_i + x_j} + \cdots + \sum_{i<j<k} \beta_{ijk}\frac{x_i x_j x_k}{(x_i + x_j + x_k)^2} + \cdots$$
$$0 \leq x_i \leq 1, \forall\, i; x_i + x_j > 0,\ \forall\, i < j. \tag{1.2.8}$$

Besides the above, there are two other homogeneous models of degree one suggested by Becker. For the sake of completeness, these are also displayed below:

$$\eta(\boldsymbol{x}) = \sum_i \beta_i x_i + \sum_{i<j} \beta_{ij}\min(x_i, x_j) + \sum_{i<j<k} \beta_{ijk}\min(x_i, x_j, x_k) + \cdots$$
$$0 < x_i < 1, \forall\, i. \tag{1.2.9}$$

$$\eta(\boldsymbol{x}) = \sum_i \beta_i x_i + \sum_{i<j} \beta_{ij}(x_i x_j)^{1/2} + \cdots + \sum_{i<j<k} \beta_{ijk}(x_i x_j x_k)^{1/3} + \cdots$$
$$0 < x_i < 1, \forall\, i. \tag{1.2.10}$$

Draper–St. John's model is given by

$$\eta(\boldsymbol{x}) = \beta_1 x_1 + \cdots + \beta_q x_q + \frac{\alpha_1}{x_1} + \cdots + \frac{\alpha_q}{x_q},\quad 0 < x_i < 1,\ \forall\, i. \tag{1.2.11}$$

Two other entirely different models will be introduced in later chapters.

1.3 Mixture Designs

Mixture designs are essentially design layouts with the descriptions of the distinct design points or vectors of the type $\boldsymbol{x}$ of mixing proportions inside the simplex, along with specification of their corresponding masses. In a sense, a mixture design is full of arbitrariness, in terms of the design points and their mass distribution. Generally, we first consider a collection of design points and then attribute a mass distribution to them. Loosely speaking, a collection of design points is also referred to as a mixture design. [The underlying mass distribution is tacitly understood to be defined at a subsequent stage, with a positive mass attributed to each design point in the collection.] In this sense, following three are the most commonly used standard mixture designs introduced by Scheffé (1958, 1963).

1.3.1 Simplex Lattice Designs

This class of designs consists of all feasible combinations of the mixing proportions wherein each proportion comprises of the values $(0, 1/m, 2/m, \ldots, m/m = 1)$ for a given integer parameter $m > 1$. Though there are $(m+1)^q$ possible combinations, only those combinations $(x_1, x_2, \ldots, x_q)$ are feasible which satisfy $x_1 + x_2 + \cdots + x_q = 1$. In a lattice design with q components and a given integer parameter m, the support set of design points is called a (q, m) simplex lattice. For a (q, m) simplex lattice design, there are $C(q + m - 1, m)$ design points where $C(a, b)$ stands for the usual binomial coefficient involving positive integers $a \geq b > 0$. .

For example, for a (4, 3) simplex lattice, the design points are given by [(1, 0, 0, 0) and its 3 variations, (1/3, 1/3, 1/3, 0) and its 3 variations, (2/3, 1/3, 0, 0) and its 11 variations]—with a total of 20 design points.

1.3.2 Simplex Centroid Designs

The centroid of a set of q nonzero coordinates in a q-dimensional coordinate system is the unique point $(1/q, 1/q, \ldots, 1/q)$. On the other hand, centroid of a set of t nonzero coordinates in a q-dimensional coordinate system is not unique. There are $C(q, t)$ centroid points of the form $[(1/t, 1/t, \ldots, 1/t, 0, 0, \ldots, 0); (1/t, 1/t, \ldots, 0, 1/t, 0, 0, \ldots, 0); \ldots\ldots (0, 0, \ldots, 0, 1/t, 1/t, \ldots, 1/t)]$.

A simplex centroid design deals exclusively with the centroids of the coordinate system, starting with exactly one nonzero component in the mixture (having q centroid points) and extending up to q nonzero components (having unique centroid point displayed above). Thus a simplex centroid design in the q-dimensional coordinate system contains $2^q - 1$ points.
As an example, for $q = 4$, there is a total of 15 points in the simplex centroid design:

[(1/4, 1/4, 1/4, 1/4); (1/3, 1/3, 1/3, 0); (1/3, 1/3, 0, 1/3); (1/3, 0, 1/3, 1/3); (0, 1/3, 1/3, 1/3); (1/2, 1/2, 0, 0); (1/2, 0, 1/2, 0); (1/2, 0, 0, 1/2); (0, 1/2, 1/2, 0); (0, 1/2, 0, 1/2); (0, 0, 1/2, 1/2); (1, 0, 0, 0); (0, 1, 0, 0); (0, 0, 1, 0); (0, 0, 0, 1)].

1.3.3 Axial Designs

It is to be noted that both simplex lattice and simplex centroid designs contain boundary points, i.e., points on the vertices, edges, and faces, *except* the centroid point $(1/q, 1/q, \ldots, 1/q)$ which lies inside the simplex. On the other hand, designs with interior points on the axis joining the points $x_i = 0, x_j = 1/(q-1), \ \forall j (\neq i)$ and $x_i = 1, x_j = 0, \ \forall j (\neq i)$ are called axial designs. Thus axial designs contain points of the form $[\{1 + (q-1)\Delta\}/q, (1-\Delta)/q, \ldots, (1-\Delta)/q]$ and their permutations, $-1/(q-1) < \Delta < 1$. These designs are essentially 'interior point' designs.

For example, taking $q = 4$ and $\Delta = 0.20$, we can form an axial design with the points $[(0.4, 0.2, 0.2, 0.2); (0.2, 0.4, 0.2, 0.2); (0.2, 0.2, 0.4, 0.2); (0.2, 0.2, 0.2, 0.4)]$.

Note that for $q = 4$, we have a choice of Δ such as $-0.33 < \Delta < 1.00$, and an axial design in general terms will be formed out of a few choices for Δ in the stated range.

In this monograph, we have dealt with such well-known mixture models and address the questions of optimal/efficient estimation of the model parameters and their meaningful functions. It turns out that the above three types of standard mixture designs occupy central stage during the investigation on optimal mixture experiments. These basic standard mixture designs will be discussed again in Chap. 3.

1.4 Exact Versus Approximate or Continuous Designs

An exact design deals with integer number of replications of the design points, thereby resulting into a design with a totality of an exact integer number of observations. For example, $[(0.4, 0.3, 0.2, 0.1), (3); (0.2, 0.4, 0.1, 0.3), (4); (0.1, 0.1, 0.4, 0.4), (2); (0.1, 0.2, 0.2, 0.5), (4)$ produces an exact design with 4 distinct design points, with repeat numbers 3, 4, 2, 4, respectively, so that altogether 13 observations are produced upon its application. On the other hand, an example of a continuous design is given by $[(0.4, 0.3, 0.2, 0.1), (0.3); (0.2, 0.4, 0.1, 0.3), (0.4); (0.1, 0.1, 0.4, 0.4), (0.2); (0.1, 0.2, 0.2, 0.5), (0.1)$. Here, again we have four distinct design points with respective 'mass' distribution given by 0.3, 0.4, 0.2, 0.1. In applications, for a given total number of observations, say $N = 30$ observations, the respective repeat numbers for the above design points are given by 9, 12, 6, 3. For N not a multiple of 10, we make nearest integer approximations as usual. The exact-design version of a continuous design has the above interpretation.

In optimality theory for 'regression models', almost exclusively, continuous design frameworks have been used. In this monograph as well, we will deal exclusively with this framework.

1.5 Applications of Mixture Methodology

Mixture experiments are commonly encountered in industrial product formulations, such as in food processing, chemical formulations, textile fibers, and pharmaceutical drugs. Some examples follow.

1. A large number of these experiments are also carried out in agriculture, where a fixed quantity of inputs such as fertilizer, irrigation water, insecticide, or pesticide

is applied as a mixture of two or more components to a crop. This makes the yield a function of the ingredient proportions.

2. In pharmaceutical drug preparation, polymers and diluents play important roles in the preparation of an inert matrix tablet, and a study of optimum mixture designs is necessary for the estimation of the parameters of the model defining the relationship between the mean response and the proportions of polymers and diluents used.
3. Intercropping is an important feature of dryland farming and has proved very useful for survival of small and marginal farmers in tropical and subtropical regions. In replacement series agricultural experiments, the component crop is introduced by replacing a part of the main crop. For fixed area under each experimental unit, the mean response is found to depend only on the proportions of the area allotted to the crops.
4. Experiments are conducted in food/horticulture technology to identify the best blending of fruit juice/pulp of lime, aonla, grape, pineapple, and mango that maximizes the responses (viz. hedonic scores on color, aroma, taste, and all taken together) among some specified mixing proportions. This presupposes establishing a relationship between the mean responses and the blending proportions, which helps to estimate the optimum blending.

1.6 Chapter-Wise Coverage of Topics

Not to obscure the flow of the chapters and the sustained interest of the readers, in Chap. 2, we make a concerted attempt to review the vast literature on optimal regression designs—admittedly much to the discontent of a serious reader—only to 'relate' to what is required for an understanding of the nature of optimal mixture experiments and the underlying optimality criteria—as discussed in this monograph! This is justified once we recognize that mixture models are effectively special types of regression models. Standard concepts of exact and approximate (or continuous) designs are well known; for the sake of completeness, these are introduced here. Next, in Chap. 3, we have introduced some of the commonly encountered mixture models and, thereafter, discussed about estimation of the underlying model parameters at length. This is done with reference to linear and quadratic homogeneous mixture models only. Some other models are deferred to latter chapters. Specific mixture designs with appealing features are also introduced in this chapter with the aim of setting the tone for the kind of optimal mixture experiments generally expected to be encountered in such studies.

Chapter 4 onwards, we deal exclusively with optimality studies in the context of mixture experiments and associated model parameters, or parametric functions with meaningful interpretations. The results are vast and varied and spread out in many directions. We progress in a manner that seemed most appealing to us in terms of the thought process of the researchers. We mention in passing that only the two standard optimality criteria [viz., A- and D-optimality] have been mostly dealt with in this monograph. In Chap. 4, we treat problems related to optimal estimation of natural parameters in mixture models due to Scheffé with reference to the unconstrained factor space [a simplex]; and in Chap. 5, we deal separately with some naturally arising constraints involving the factor space. Chapter 6 is meant for discussions on natural parameters in other mixture models. In the next four chapters (Chaps. 7–10), we discuss at length the problems associated with optimal estimation of some nonlinear functions of the model parameters. These functions arise naturally while one tries to maximize the 'expected output' as per the model specifications. Scheffé model, Darroch–Waller and Log-contrast models are taken up in these chapters. Lastly, we discuss some applications in Chap. 11 and a few miscellaneous diverse topics in Chap. 12. The topics are: robust mixture designs, optimality in Scheffé's and Darroch–Waller models with random regression coefficients, optimality in mixture–amount model, multi-response mixture models, and mixture designs in blocks.

References

Becker, N. G. (1968). Models for the response of a mixture. *Journal of the Royal Statistical Society Series B (Methodological)*, *30*, 349–358.

Cornell, J. A. (2002). *Experiments with Mixtures* (3rd ed.). New York: Wiley.

Draper, N. R., & St. John, R. C. (1977). A mixture model with inverse terms. *Technometrics*, *19*, 37–46.

Scheffé, H. (1958). Experiments with mixtures. *Journal of the Royal Statistical Society Series B (Methodological)*, *20*, 344–360.

Scheffé, H. (1963). Simplex-centroid design for experiments with mixtures. *Journal of the Royal Statistical Society Series B (Methodological)*, *25*, 235–263.

Chapter 2
Optimal Regression Designs

Abstract In this chapter, we review the theory of optimum regression designs. Concept of continuous design and different optimality criteria are introduced. The role of *de la Garza phenomenon* and *Loewner order domination* are discussed. Equivalence theorems for different optimality criteria, which play an important role in checking the optimality of a given otherwise prospective design, are presented. These results are repeatedly used in later chapters in the search for optimal mixture designs. We present standard optimality results for single variable polynomial regression model and multivariate linear and quadratic regression model. Kronecker product representation of the model(s) and related optimality results are also discussed.

Keywords Continuous design · Optimality criteria · de la Garza phenomenon · Loewner order domination · Polynomial regression models · Equivalence theorem · Optimum regression designs

2.1 Introduction

In this chapter, we will discuss optimality aspects of regression designs in an *approximate* (or, *continuous*) design setting defined below.

Let y be the observed response at a point $(x_1, x_2, \ldots, x_k) = \boldsymbol{x}$ varying in some k-dimensional experimental domain $\mathcal{X}$ following the general linear model

$$y(\boldsymbol{x}) = \eta(\boldsymbol{x}, \boldsymbol{\beta}) + e(\boldsymbol{x}), \tag{2.1.1}$$

with usual assumptions on error component $e(\boldsymbol{x})$, viz. mean zero and uncorrelated homoscedastic variance σ^2; $\eta(\boldsymbol{x}, \boldsymbol{\beta})$ is the mean response function involving k or more unknown parameters. Once for all, we mention that $\boldsymbol{x}$ will represent a combination of the mixing components, the number of such components will be understood

B. K. Sinha et al., *Optimal Mixture Experiments*, Lecture Notes in Statistics 1028,
DOI: 10.1007/978-81-322-1786-2_2, © Springer India 2014

from the context. Moreover, the same will be used to denote a row or a column vector, as the context demands.

Generally, it is assumed that in the region of immediate interest, $\eta(\boldsymbol{x}, \boldsymbol{\beta})$ can be approximated by a polynomial of certain degree and can be expressed as

$$\eta(\boldsymbol{x}, \boldsymbol{\beta}) = f'(\boldsymbol{x})\boldsymbol{\beta}. \tag{2.1.2}$$

The discrete or, exact designing problem is that of choosing N design points in the experimental domain $\mathcal{X}$ so that individually each of the t parameters of the mean response function can be estimated with satisfactory degree of accuracy. A *continuous* or an *approximate design* ξ for model (2.1.2), as introduced by Kiefer (1959), consists of finitely many distinct *support points* $\boldsymbol{x}_1, \boldsymbol{x}_2, \ldots, \boldsymbol{x}_n \in \mathcal{X}$, at which observations of the response are to be taken, and of corresponding design weights $\xi(\boldsymbol{x}_i) = p_i, i = 1, 2, \ldots, n$ which are positive real numbers summing up to 1. In other words, an approximate design ξ is a probability distribution with finite support on the factor space $\mathcal{X}$ and is represented by

$$\xi = \{\boldsymbol{x}_1, \boldsymbol{x}_2, \ldots, \boldsymbol{x}_n; p_1, p_2, \ldots, p_n\}, \tag{2.1.3}$$

which assigns, respectively, masses $p_1, p_2, \ldots, p_n; p_i > 0, \sum p_i = 1$, to the n distinct support points $\boldsymbol{x}_1, \boldsymbol{x}_2, \ldots, \boldsymbol{x}_n$ of the design ξ in the experimental region [may be a subspace of the factor space $\mathcal{X}$]. Let $\mathcal{D}$ be the class of all competing designs. For a given N, a design ξ cannot, in general, be properly realized, unless its weights are integer multiples of $1/N$ i.e., unless $n_i = Np_i, i = 1, 2, \ldots, n$ are integers with $\sum n_i = N$. An approximate design becomes an *exact design* of size N in the special case when $n_i = Np_i, i = 1, 2, \ldots, n$ are integers.

The *information matrix* (often termed 'per observation moment matrix') for β, using a design ξ, is given by

$$M(\xi) = \sum p_i f(\boldsymbol{x}_i) f'(\boldsymbol{x}_i). \tag{2.1.4}$$

It may be noted that for unbiased estimation of the parameters in the mean response function, it is necessary that the number of 'support points' i.e., $\boldsymbol{x}_i$s must be at least 't,' the number of model parameters. It is tacitly assumed that the choice of a design leads to unbiased estimation of the parameters and as such the information matrix M of order $t \times t$ is a positive definite matrix. Let $\mathcal{M}$ denote the class of all positive definite moment matrices. As we will see in the rest of this monograph, the information matrix (2.1.4) of a design plays an important role in the determination of an optimum design. In fact, most of the optimality criteria are different functions of the information matrix.

2.2 Optimality Criteria

The utility of an optimum experimental design lies in the fact that it provides a design ξ^* that is best in some sense. Toward this let us bring in the concept of *Loewner ordering*. A design ξ_1 dominates another design ξ_2 in the Loewner sense if $M(\xi_1) - M(\xi_2)$ is a nonnegative definite (nnd) matrix. Thus, Loewner partial ordering among information matrices induces a partial ordering among the associated designs. We shall denote $\xi_1 \succ \xi_2$ when ξ_1 dominates ξ_2 in the Loewner sense. A design ξ^* that dominates over all other designs in $\mathcal{D}$ in the Loewner sense is called *Loewner optimal*. In general, there exists no Loewner optimal design ξ^* that dominates every other design ξ in $\mathcal{D}$ [vide Pukelsheim (1993)]. A popular way out is to specify an optimality criterion, defined as a real-valued function of $M(\xi)$. An optimal design is one whose moment matrix *minimizes* the criterion function $\phi(\xi)$ over a well-defined set of competing moment matrices (or designs); vide Shah and Sinha (1989) and Pukelsheim (1993) for details. Let $0 < \lambda_1 \leq \lambda_2 \leq \cdots \leq \lambda_t$ be the t positive characteristic roots of the moment matrix M. It is essential that a reasonable criterion ϕ conforms to the Loewner ordering

$$M(\xi_1) \geq M(\xi_2) \Rightarrow \phi(M(\xi_1)) \geq \phi(M(\xi_2)).$$

The first original contribution on optimum regression design is by Smith (1918) who determined the G-optimum design for the estimation of parameters of a univariate polynomial response function. After a gap of almost four decades, a number of contributions in this area were made by, Elfving (1952), Chernoff (1953), Ehrenfield (1955), Guest (1958), Hoel (1958). Extending their results, Kiefer (1958, 1959, 1961), Kiefer and Wolfowitz (1959) developed a systematic account of different optimality criteria and related designs. These can be discussed in terms of maximizing the function $\phi(M(\xi))$ of $M(\xi)$.

The most prominent optimality criteria are the *matrix means* ϕ_p, for $p \in (-\infty, 2]$, which enjoy many desirable properties. These were introduced by Kiefer (1974, 1975):

$$\phi_p(M) = \left(\left(\frac{1}{t}\right)\sum_i \lambda_i^p\right)^{1/p}$$

where $\lambda_1, \lambda_2, \ldots, \lambda_t$ denote the eigenvalues of the positive definite information matrix $M(\xi)$ of order $t \times t$. Excluding trivial cases, it is evident that an optimum design which satisfies all the criteria does not exist. The classical A-, D- *and E-optimality criteria* are special cases of ϕ_p-optimality criteria. The criterion $\phi_{-1}(M)$ is the A-optimality criterion. Maximizing $\phi_{-1}(M)$ is equivalent to minimizing the trace of the corresponding dispersion matrix (in the exact or asymptotic sense). The D-optimality criterion $\phi_0(M)$ is equivalent to maximizing the determinant det.(M). The extreme member of $\phi_p(M)$ for $p \to -\infty$ yields the smallest eigen (E-optimality criterion) $\phi_{-\infty}(M) = \lambda_{\min}(M)$. Besides, there are other optimality criteria for comparing designs viz. G-optimality criterion, D_s- and D_A-optimality criteria, I-optimality criterion etc. [For different optimality criteria and their statistical significance, the readers are referred to Fedorov (1972), Silvey (1980), Shah and Sinha (1989), Pukelsheim (1993)].

In general, the direct search for optimum design may be prohibitive. The degree of difficulty depends on the nature of response function, criterion function and/or the experimental region. However, there are tools that can be used to reduce, sometimes substantially, the class of competing designs.

2.3 One Dimensional Polynomial Regression

In practice, polynomial models are widely used because they are flexible and usually provide a reasonable approximation to the true relationship among the variables. Polynomial models with low order, whenever possible, are generally recommended. Higher order polynomial may provide a better fit to the data and hence an improved approximation to the true relationship; but the numerous coefficients in such models make them difficult to interpret. Sometimes, polynomial models are used after an appropriate transformation has been applied on the independent variables to lessen the degree of nonlinearity. Examples of such transformations are the logarithm and square transformations. Box and Cox (1964), Carroll and Ruppert (1984) gave a detailed discussion on the use and properties of various transformations for improving fit in linear regression models.

In a one-dimensional polynomial regression, the mean response function is given by

$$\eta(x) = f'(x)\boldsymbol{\beta} = \beta_0 + \beta_1 x + \beta_2 x^2 + \cdots + \beta_d x^d, \tag{2.3.1}$$

where $f'(x) = (1, x, x^2, \ldots, x^d)$ and $\boldsymbol{\beta}' = (\beta_0, \beta_1, \ldots, \beta_d)$. Several authors attempted to find optimum designs for the estimation of parameters of the above model. As mentioned earlier, Smith (1918) first obtained G-optimum designs for the estimation of parameters. de la Garza (1954) considered the estimation of parameters of the above model from N observations in a given range. By changing the origin and scale, the domain of experimental region, i.e., the factor space may be taken to be $\mathcal{X} = [-1, 1]$. Consider a design ξ given by (2.1.3) in the factor space $[-1, 1]$ with information matrix (2.1.4). de la Garza (1954) showed that corresponding to any arbitrary continuous design ξ as in (2.1.3) supported by n [$>d + 1$] distinct points, there exists a design with exactly $d + 1$ support points such as

$$\xi^* = \{x_1^*, x_2^*, \ldots, x_{d+1}^*; p_1^*, p_2^*, \ldots, p_{d+1}^*\}, \tag{2.3.2}$$

for which the information matrices are the same, i.e, $M(\xi) = M(\xi^*)$. Moreover, $x_{\min} \leq x_{\min}^* \leq x_{\max}^* \leq x_{\max}$. This appealing feature of the two designs is referred to as *information equivalence*. Afterward, the de la Garza phenomenon has been extensively studied by Liski et al. (2002), Dette and Melas (2011), Yang (2010).

In addition to this, Pukelsheim (1993) extensively studied the phenomenon of *information domination* in this context.

Remark 2.3.1 The exact design analog of the feature of information equivalence is generally hard to realize. Mandal et al. (2014) provide some initial results in this direction.

In general, different optimal designs may require different number of design points. It is clear that in order to estimate 't' parameters in any model, at least 't' distinct design points are needed, and for many models and optimality criteria, the optimal number of distinct design points will be 't.' For nonlinear models, the information matrix has an interpretation in an approximate sense, as being the inverse of the asymptotic variance-covariance matrix of the estimates of the model parameters. An interesting result called *Caratheodory's Theorem* provides us with an upper bound on the number of design points needed for the existence of a positive definite information matrix. For many design problems with 't' parameters, this number is '$t(t + 1)/2$.' Thus the optimal number of distinct design points is between 't' and '$t(t + 1)/2$.' Finally, it should be noted that the upper bound does not hold for the *Bayesian* design criteria (Atkinson et al. 2007, Chap. 18).

Guest (1958) obtained general formulae for the distribution of the points of observations and for the variances of the fitted values in the minimax variance case, and compared the variances with those for the uniform spacing case. He showed that the values of $x_1, x_2, \ldots, x_{d+1}$, (with reference to the model (2.3.1)) that minimize the maximum variance of a single estimated ordinate are given by means of the zeros of the derivative of a Legendre polynomial. Hoel (1958) used the D-optimality criterion for determining the best choice of fixed variable values within an interval for estimating the coefficients of a polynomial regression curve of given degree for the classical regression model. Using the same criterion, some results are obtained on the increased efficiency arising from doubling the number of equally spaced observation points (i) when the total interval is fixed and (ii) when the total interval is doubled. Measures of the increased efficiency are found for the classical regression model and for models based on a particular stationary stochastic process and a pure birth stochastic process. Moreover, he first noticed that D- and G-optimum designs *coincide* in a one-dimensional polynomial regression model.

Kiefer and Wolfowitz (1960) extended and established this phenomenon of coincidence to any linear model through what is now known as 'Equivalence Theorem.' Writing $d(\boldsymbol{x}, \xi) = f'(\boldsymbol{x})M^{-1}(\boldsymbol{\beta}, \xi)f(\boldsymbol{x})$, the celebrated equivalence theorem of Kiefer and Wolfowitz (1960) can be stated as follows:

Theorem 2.3.1 *The following assertions:*

(i) the design ξ^* *minimizes* $| M^{-1}(\boldsymbol{\beta}, \xi) |$,
(ii) the design ξ^* *minimizes* $\max_{\boldsymbol{x}} d(\boldsymbol{x}, \xi)$,
(iii) $\max_{\boldsymbol{x}} d(\boldsymbol{x}, \xi^*) = t$

are equivalent. The information matrices of all designs satisfying (i)–(iii) coincide among themselves. Any linear combination of designs satisfying (i)–(iii) also satisfies (i)–(iii).

In this context, Fedorov (1972) also serves as a useful reference. This theorem plays an important role in establishing the D-optimality of a given design obtained from intuition or otherwise. Moreover, it gives the nature of the support points of an optimum design. For example, let us consider a quadratic regression given by

$$y_i = \beta_0 + \beta_1 x_i + \beta_2 x_i^2 + e_i, i = 1, 2, \ldots, n \tag{2.3.3}$$

where β_i s are fixed regression parameters and $e_i s$ are independent random error with usual assumptions, viz. mean 0 and variance σ^2. We assume, as before, that the factor space is $\mathcal{X} = [-1, 1]$.

The information matrix for an arbitrary design $\xi = \{x_1, x_2, \ldots, x_n; p_1, p_2, \ldots, p_n\}$ can be readily written down, and it involves the moments of x-distribution, i.e., $\mu'_r = \sum p_i x_i^r; r = 1, 2, 3, 4$. It is well known that the information matrix M is positive definite iff $n > 2$, since the $x_i s$ are assumed to be all distinct without any loss of generality and essentially we are restricting to this class of designs. It is clearly seen that $d(x, \xi) = f'(x)M^{-1}(\boldsymbol{\beta}, \xi) f(x)$, with $f(x) = (1, x, x^2)'$ is *quartic in* x so that the three support points of the D-optimum design are at the two extreme points ± 1 and a point lying in between. Since the D-optimality criterion, for the present problem, is invariant with respect to sign changes, the interior support point must be at 0. Thus, the three support points of the D-optimum design in $[-1, 1]$ are 0 and ± 1. The weights at the support points are all equal since here the number of support points equals the number of parameters. It may be noted that for D-optimality, whenever the number of support points is equal to the number of parameters, the weights at the support points are necessarily equal.

Remark 2.3.2 The above result can be derived using altogether different arguments. In view of de la Garza phenomenon, given the design ξ with $n > 3$, there exists a three-point design $\xi^* = \{(a, P), (b, Q), (c, R)\}$, where $-1 \leq a < b < c \leq 1$, and $0 < P, Q, R < 1$, $P + Q + R = 1$, such that $M(\xi) = M(\xi^*)$. Again, referring to Liski et al. (2002), we may further 'improve' ξ^* to $\xi^{**} = \{(-1, P), (c, Q), (1, P)\}$ by proper choice of 'c' in the sense of Loewner Domination. It now follows that for D-optimality, det.$M(\xi^{**}) \leq (4/27)(1 - c^2)^2 \leq 4/27$ with '=' if and only if $c = 0$ and $P = Q = R = 1/3$.

Atwood (1969) observed that in several classes of problems an optimal design for estimating all the parameters is supported only on certain points of symmetry. Moreover he considered the optimality when nuisance parameters are present and obtained a new sufficient condition for optimality. He corrected a version of the condition which Karlin and Studden (1966) stated as equivalent to optimality, and proved the natural invariance theorem involving this condition. He applied these results to the problem of multi-linear regression on the simplex (introduced by Scheffé 1958) when estimating all or only some of the parameters. This will be discussed in detail in Chaps. 4–6.

Fedorov (1971, 1972) developed the equivalence theorem for *Linear optimality* criterion in the lines of equivalence theorem of Kiefer and Wolfowitz (1960) for

D-optimality. Assuming estimability of the model parameters, we denote the dispersion matrix by $D(\xi)$. It is known that $D(\xi) = M^{-1}(\xi)$. A design ξ^* is said to be linear optimal if it minimizes $L(D(\xi))$ over all ξ in Ξ where L is a linear optimality functional defined on the dispersion matrices satisfying

$$L(A + B) = L(A) + L(B)$$

for any two nnd matrices A and B and

$$L(cA) = cL(A)$$

for any scalar $c > 0$. Then, the equivalence theorem for linear optimality can be stated as follows.

Theorem 2.3.2 *The following assertions:*

(i) the design ξ^* *minimizes* $L[D(\xi)]$,
(ii) the design ξ^* *minimizes* $\max_{\boldsymbol{x}} L[D(\xi) f(\boldsymbol{x}) f'(\boldsymbol{x}) D(\xi)]$,
(iii) $\max_{\boldsymbol{x}} L[D(\xi^*) f(\boldsymbol{x}) f'(\boldsymbol{x}) D(\xi^*)] = L[D(\xi^*)]$

are equivalent. Any linear combination of designs satisfying (i)–(iii) also satisfies (i)–(iii).

Similar equivalence theorems are also available for E-optimality criterion (cf. Pukelsheim 1993).

Afterword, Kiefer (1974) introduced the ϕ-optimality criterion, a real-valued concave function defined on a set $\mathcal{M}$ of positive definite matrices. He then established the following equivalence theorem [cf. Silvey (1980), Whittle (1973)].

Let $\mathcal{M}$ be the class of all moment matrices obtained by varying design ξ in Ξ and ϕ is a real-valued function defined on $\mathcal{M}$. Then the Frechét derivative of ϕ at M_1 in the direction of M_2 is defined as

$$F_{\phi}(M_1, M_2) = \lim_{\alpha \to 0+} \frac{1}{\alpha}[\phi\{(1 - \alpha)M_1 + \alpha M_2\} - \phi(M_1)].$$

Theorem 2.3.3 *When* ϕ *is concave on* $\mathcal{M}$, ξ^* *is* ϕ*-optimal if and only if*

$$F_{\phi}(M(\xi^*), M(\xi)) \leq 0 \tag{2.3.4}$$

for all $\xi \in \mathcal{D}$.

This theorem states simply that we are at the top of a *concave mountain* when there is no direction in which we can look forward to another point on the mountain. However, since it is difficult to check (2.3.4) for all $\xi \in \mathcal{D}$, Kiefer (1974) established the following theorem that is more useful in verifying the optimality or non-optimality of a design ξ^*.

Theorem 2.3.4 *When ϕ is concave on $\mathcal{M}$ and differentiable at $M(\xi^*)$, ξ^* is ϕ-optimal if and only if*

$$F_\phi\{M(\xi^*), f(\boldsymbol{x})f'(\boldsymbol{x})\} \leq 0 \tag{2.3.5}$$

for all $\boldsymbol{x} \in \mathcal{X}$.

For the proof and other details, one can go through Kiefer (1974) and Silvey (1980). This result has great practical relevance because in many situations, the optimality problems may not be of the classical A-, D- or E-optimality type but fall under a wide class of ϕ-optimality criteria. The equivalence theorem above then helps us to establish the optimality of a design obtained intuitively or otherwise.

The equivalence theorem in some form or the other has been repeatedly used in subsequent chapters of this monograph. For the equivalence theorem for Loewner optimality or other specific optimality criteria, the readers are referred to Pukelsheim (1993).

Remark 2.3.3 An altogether different optimality criterion was suggested in Sinha (1970). Whereas all the traditional optimality criteria are exclusively functions of the (positive) eigenvalues of the information matrix, this one was an exception. In the late 1990s, there was a revival of research interest in this optimality criterion, termed as 'Distance Optimality criterion' or, simply, 'DS-optimality' criterion.

In the context of a very general linear model set-up involving a (sub)set of parameters θ admitting best linear unbiased estimator (blue) $\hat{\theta}$, it is desirable that the 'stochastic distance' between θ and $\hat{\theta}$ be the least. This is expressed by stating that the 'coverage probability'

$$\Pr[\|\hat{\theta} - \theta\| < \epsilon]$$

should be as high as possible, for every $\epsilon > 0$. As an optimal design criterion, therefore, we seek to characterize a design ξ_0 such that for every given ε, $\hat{\theta}$ based on ξ_0 provides largest coverage probability than any other competing ξ.

Sinha (1970) initiated study of DS-optimal designs for one-way and two-way analysis of variance (ANOVA) setup. Much later, the study was further continued in ANOVA and regression setup (Liski et al. 1998; Saharay and Bhandari 2003; Mandal et al. 2000). On the other hand, theoretical properties of this criterion function were studied in depth in a series of papers (Liski et al. 1999; Zaigraev and Liski 2001, 2006; Zaigraev 2005)

We will not pursue this criterion in the present Monograph.

Since in a mixture experiment, we will be concerned with a number of components, let us first review some results in the context of multi-factor experiment.

2.4 Multi-factor First Degree Polynomial Fit Models

Let us first consider a k-factor first degree polynomial fit model with no constant term, viz.,

$$y_{ij} = \beta_1 x_{i1} + \beta_2 x_{i2} + \cdots + \beta_k x_{ik} + e_{ij}, \tag{2.4.1}$$

with k regressor variables, n experimental conditions $\boldsymbol{x_i} = (x_{i1}, x_{i2}, \ldots, x_{ik})$, $i = 1, 2, \ldots, n$; $j = 1, 2, \ldots, N_i$, $\sum N_i = N$. Most often we deal with a continuous or approximate theory version of the above formulation in which $p_i s$ are regarded as (positive) 'mass' attached to the points $\boldsymbol{x}_i$s, subject to the condition $\sum p_i = 1$.

In polynomial fit model with single factor, the experimental domain is generally taken as $\mathcal{X} = [-1, +1]$. For k-factor polynomial linear fit model (2.4.1), the experimental domain is a subset of the k-dimensional Euclidean space R^k. Generally, optimum designs are developed for the following two extensions of the one-dimensional domain $\mathcal{X} = [-1, +1]$: A *Euclidean ball* of radius $\sqrt{k}$ and a symmetric k-dimensional *hypercube* $[-1, +1]^k$. In practice, there may be other types of domains viz., a constrained region of the type $\mathcal{X}_R = [0 \leq x_i \leq 1, \sum x_i = \alpha \leq 1]$. The mixture experiment, the optimality aspect of which will be considered in details in subsequent chapters, has domain that corresponds to $\alpha = 1$.

Below we develop the continuous design theory for the above model. Consider the experimental domain for the model (2.4.1), which is a k-dimensional ball of radius $\sqrt{k}$, that is, $\mathcal{X}(k) = [\boldsymbol{x} \in R_k, \|\boldsymbol{x}\| \leq \sqrt{k}\}$, where $\|.\|$ denotes the Euclidean norm. Set $\mu_{jm} = \sum_i p_i x_{ij} x_{im}$ for $j, m = 1, 2, \ldots, k$. This has the simple interpretation as the 'product moment' of jth and mth factors in the experiment. Then, the information matrix for an n-point $(n \geq k)$ design

$$\xi = \{\boldsymbol{x}_1, \boldsymbol{x}_2, \ldots, \boldsymbol{x}_n; p_1, p_2, \ldots, p_n\}$$

is of the form

$$M(\xi) = \sum p_i f(\boldsymbol{x}_i) f'(\boldsymbol{x}_i) = \begin{pmatrix} \mu_{11} & \mu_{12} & \ldots & \mu_{1k} \\ & \mu_{22} & \ldots & \mu_{2k} \\ & & \ldots & \ldots \\ & & & \mu_{kk} \end{pmatrix} \tag{2.4.2}$$

with $f'(\boldsymbol{x}_i) = (x_{i1}, x_{i2}, \ldots, x_{ik})$.

Using spectral decomposition of the matrix $M(\xi)$, it can be easily shown that $M(\xi)$ can equivalently be represented by a design ξ^* with k orthogonal support points in $\mathcal{X}(k)$:

$$\xi^* = \{\boldsymbol{x}_1^*, \boldsymbol{x}_2^*, \ldots, \boldsymbol{x}_k^*; p_1^*, p_2^*, \ldots, p_k^*\} \tag{2.4.3}$$

i.e., $M(\xi^*) = M(\xi)$. Such a design is termed as orthogonal design (Liski et al. 2002). This incidentally demonstrates validity of the de la Garza phenomenon (DLG phenomenon) in the multivariate linear setup without the constant term. We can further improve over this design in terms of the Loewner order domination of the information matrix by stretching the mass at the boundary of $\mathcal{X}(k)$. In other words, given an orthogonal design ξ^* as in (2.4.3), there exists another k-point orthogonal design

$$\xi^{**} = \{\boldsymbol{x}_1^{**}, \boldsymbol{x}_2^{**}, \ldots, \boldsymbol{x}_k^{**};\, p_1^*, p_2^*, \ldots, p_k^*\} \tag{2.4.4}$$

with $\boldsymbol{x}_i^{**} = \sqrt{k}\boldsymbol{x}_i^*/\|\boldsymbol{x}_i^*\|$, such that $M(\xi^{**}) - M(\xi^*)$ is nnd i.e., $\xi^{**} \succ \xi^* \sim \xi$. One can now determine optimum designs in the class of designs (2.4.4) using different optimality criteria. Similar results hold for multi-factor linear model with constant term.

A symmetric k-dimensional unit cube $[-1, +1]^k$ is a natural extension of $[-1, +1]$. Note that $[-1, +1]^k$ is the convex hull of its extreme points, the 2^k vertices of $[-1, +1]^k$. It is known that in order to find optimal support points, we need to search the extreme points of the regression range only. If the support of a design contains other than extreme points, then it can be Loewner dominated by a design with extreme support points only. This result was basically presented by Elfving (1952, 1959). A unified general theory is given by Pukelsheim (1993).

A generalization of the model (2.4.1) incorporating the constant term has been studied in Liski et al. (2002). Also details for the latter factor space described above have been given there. We do not pursue these details here.

In the context of mixture models, as has been indicated before, we do not include a constant term in the mean model. So, the above study may have direct relevance to optimality issues in mixture models.

2.5 Multi-factor Second Degree Polynomial Fit Models

Consider now a second-degree polynomial model in k variables:

$$\eta_x = \beta_0 + \sum_{i=1}^{k} \beta_i x_i + \sum_{i=1}^{k} \sum_{j>i}^{k} \beta_{ij} x_i x_j. \tag{2.5.1}$$

For the second-degree model, in finding optimum designs, it is more convenient to work with the Kronecker product representation of the model (cf. Pukelsheim 1993).

For a k-factor second-degree model, $k \geq 2$, let us take the regression function to be

$$\eta(\boldsymbol{x}, \boldsymbol{\beta}) = g'(\boldsymbol{x})\boldsymbol{\beta} \tag{2.5.2}$$

where

$$g'(\boldsymbol{x}) = (1, \boldsymbol{x}', \boldsymbol{x}' \otimes \boldsymbol{x}'), \tag{2.5.3}$$

$\boldsymbol{\beta}$ is a vector of parameters and the factor space is given by

$$\mathcal{X}(k) = \{\boldsymbol{x} : \|\boldsymbol{x}\| \leq k\}.$$

To characterize the optimum design for the estimation of $\boldsymbol{\beta}$, let us consider the following designs:

$$\begin{aligned}\xi_0 &= \{\boldsymbol{x} \mid \boldsymbol{x}'\boldsymbol{x} = 0\}\\ \xi_c &= \frac{1}{2^r} \text{ fraction of a } 2^k \text{ factorial experiment with levels } \pm 1\\ \xi_s &= \text{ set of star points of the form } (\pm\sqrt{k}, (0, 0, \ldots, 0), (0, \pm\sqrt{k}, 0, \ldots, 0), \ldots,\\ &\quad (0, 0, \ldots, \pm\sqrt{k})\\ \tilde{\xi}_{\sqrt{k}} &= \frac{n_c\xi_c + n_s\xi_s}{n}, n_c = 2^{k-r}, n_s = 2k, n = n_c + n_s.\end{aligned}$$

A design $\xi^* = (1-\alpha)\xi_0 + \alpha\,\tilde{\xi}_{\sqrt{k}}$ is called a *central composite design* (CCD) (cf. Box and Wilson 1951). Such a design ξ^* is completely characterized by α. It is understood that $0 \leq \alpha \leq 1$.

Before citing any result on optimum design in the second-order case, let us first of all bring in the concept of Kiefer optimality. Symmetry and balance have always been a prime attribute of good experimental designs and comprise the first step of the Kiefer design ordering. The second step concerns the usual Loewner matrix ordering. In view of the symmetrization step, it suffices to search for improvement when the Loewner ordering is restricted only to exchangeable moment matrices.

Now, we cite a very powerful result on Kiefer optimality in the second-order model (2.5.1).

Theorem 2.5.1 *The class of CCD is complete in the sense that, given any design, there is always a CCD that is better in terms of*

(i) Kiefer ordering
(ii) ϕ-optimality, provided it is invariant with respect to orthogonal transformation.

There are many results for specific optimality criteria for the second-order model (see e.g., Pukelsheim 1993). We are not going to discuss the details.

It must be noted that in the context of mixture models, we drop the constant term β_0 from the mean model. Moreover, the factor space (constrained or not) is quite different from unit ball/unit cube. Yet, the approach indicated above has been found to be extremely useful in the characterization of optimal mixture designs. All these will be discussed in details from Chap. 4 onward.

References

Atkinson, A. C., Donev, A. N., & Tobias, R. D. (2007). *Optimum experimental designs, in SAS*. Oxford: Oxford University Press.

Atwood, C. L. (1969). Optimal and efficient designs of experiments. *The Annals of Mathematical Statistics, 40*, 1570–1602.

Box, G. E. P., & Wilson, K. B. (1951). On the experimental attainment of optimum conditions. Journal of the Royal Statistical Society. *Series B (Methodological)*, *13*, 1–38.

Box, G. E. P., & Cox, D. R. (1964). An analysis of transformations. *Journal of the Royal Statistical Society, Series B*, *26*, 211–252.

Carroll, R. J., & Ruppert, D. (1984). Power transformations when fitting theoretical models to data. *Journal of the American Statistical Association*, *79*, 321–615.

Chernoff, H. (1953). Locally optimal designs for estimating parameters. *The Annals of Mathematical Statistics*, *24*, 586–602.

Dette, H., & Melas, V. B. (2011). A note on the de la Garza phenomenon for locally optimum designs. *The Annals of Statistics*, *39*, 1266–1281.

Ehrenfield, S. (1955). On the efficiency of experimental designs. *The Annals of Mathematical Statistics*, *26*, 247–255.

Elfving, G. (1959). Design of linear experiments. In U. Grenander (Ed.), Probability and statistics. The Herald cramer volume (pp. 58–74). Stockholm: Almquist and Wiksell.

Elfving, G. (1952). Optimum allocation in linear regression theory. *The Annals of Mathematical Statistics*, *23*, 255–262.

Fedorov, V. V. (1971). Design of experiments for linear optimality criteria. *Theory of Probability & its Applications*, *16*, 189–195.

Fedorov, V. V. (1972). *Theory of optimal experiments*. New York: Academic.

de la Garza, A. (1954). Spacing of information in polynomial regression. *The Annals of Mathematical Statistics*, *25*, 123–130.

Guest, P. G. (1958). The spacing of observations in polynomial regression. *The Annals of Mathematical Statistics*, *29*, 294–299.

Hoel, P. G. (1958). The efficiency problem in polynomial regression. *The Annals of Mathematical Statistics*, *29*, 1134–1145.

Karlin, S., & Studden, W. J. (1966). Optimal experimental designs. *The Annals of Mathematical Statistics*, *37*, 783–815.

Kiefer, J. (1958). On non-randomized optimality and randomized non-optimality of symmetrical designs. *The Annals of Mathematical Statistics*, *29*, 675–699.

Kiefer, J. (1959). Optimum experimental designs. Journal of the Royal Statistical Society. *Series B (Methodological)*, *21*, 272–304.

Kiefer, J., & Wolfowitz, J. (1959). Optimum designs in regression problems. *The Annals of Mathematical Statistics*, *30*, 271–294.

Kiefer, J., & Wolfowitz, J. (1960). The equivalence of two extremum problems. *Canadian Journal of Mathematics*, *12*, 363–366.

Kiefer, J. (1961). Optimum designs in regression problems II. *The Annals of Mathematical Statistics*, *32*, 298–325.

Kiefer, J. (1974). General equivalence theory for optimum designs (approximate theory). *The annals of Statistics*, *2*, 849–879.

Kiefer, J. (1975). Optimum design: Variation in structure and performance under change of criterion. *Biometrika*, *62*, 277–288.

Liski, E. P., Luoma, A., Mandal, N. K., & Sinha, B. K. (1998). Pitman nearness, distance criterion and optimal regression designs. *Bulletin of the Calcutta Statistical Association*, *48*, 179–194.

Liski, E. P., Luoma, A., & Zaigraev, A. (1999). Distance optimality design criterion in linear models. *Metrika*, *49*, 193–211.

Liski, E. P., Mandal, N. K., Shah, K. R., & Sinha, B. K. (2002). *Topics in optimal design* (Vol. 163). Lecture notes in statistics New York: Springer.

Mandal, N. K., Pal, M., & Sinha, B. K. (2014). Some finer aspects of the de la Garza phenomenon : A study of exact designs in linear and quadratic regression models. To appear in *Statistics and Applications* (special volume in the memory of Prof. M.N. Das).

Mandal, N. K., Shah, K. R., & Sinha, B. K. (2000). Comparison of test vs. control treatments using distance optimality criterion. *Metrika*, *52*, 147–162.

Pukelsheim, F. (1993). *Optimal design of experiments*. New York: Wiley.

Saharay, R., & Bhandari, S. K. (2003). D_S-optimal designs in one-way ANOVA. *Metrika, 57*, 115–125.
Scheffé, H. (1958). Experiments with mixtures. Journal of the Royal Statistical Society. *Series B (Methodological), 20*, 344–360.
Shah, K. R., & Sinha, B. K. (1989). *Theory of optimal designs* (Vol. 54). Lecture notes in statistics New York: Springer.
Silvey, S. D. (1980). *Optimal designs*. London: Chapman Hall.
Sinha, B. K. (1970). On the optimality of some designs. *Calcutta Statistical Association Bulletin, 20*, 1–20.
Smith, K. (1918). On the standard deviations of adjusted and interpolated values of an observed polynomial function and its constants and the guidance they give towards a proper choice of the distribution of observations. *Biometrika, 12*, 1–85.
Whittle, P. (1973). Some general points in the theory of optimal experimental design. Journal of the Royal Statistical Society. *Series B (Methodological), 35*, 123–130.
Yang, M. (2010). On the de la Garza phenomenon. *The Annals of Statistics, 38*, 2499–2524.
Zaigraev, A., & Liski, E. P. (2001). A stochastic characterization of Loewner optimality design criterion in linear models. *Metrika, 53*, 207–222.
Zaigraev, A. (2005). Stochastic design criteria in linear models. *Austrian Journal of Statistics, 34*, 211–223.
Zaigraev, A., & Liski, E. P. (2006). On D_S-optimal design matrices with restrictions on rows or columns. *Metrika, 64*, 181–189.

Chapter 3
Parameter Estimation in Linear and Quadratic Mixture Models

Abstract In this chapter, we present standard mixture models and standard mixture designs as generally applied to such models. These mixture models and mixture designs will occur throughout the monograph. Several generalizations of standard mixture designs are also discussed here. Estimability issues involving the model parameters are addressed at length. In the process, information matrices are worked out and their roles are emphasized. The concept of Loewner domination is also brought in.

Keywords Scheffé's linear and quadratic mixture models · Becker's homogeneous model of degree one · Draper–St. John's model · Parameter estimation · Simplex lattice design · Simplex centroid design · Axial design · Generalized axial designs

3.1 Introduction

Mixture experiments were first discussed in Quenouille (1953). Later on, Scheffé (1958, 1963) made a systematic study and laid a strong foundation. A comprehensive study up to modern developments can be found in Cornell (2002). Mixture experiments deal with typical multiple regression models, wherein the regressors are the proportions of the mixing components. Thus, mixture experiment is useful in the study of the quality of products like polymers, paint, concrete, alloys, glass, etc., which depend on the relative proportions of the ingredients in the products. Examples of mixture experiments are also found in the pharmaceutical industry and the food industry; vide Chap. 11.

There is a slight overlap of this chapter with Chap. 1. Sections 3.1 and 3.2 are mostly repetitions of what have been presented in Chap. 1. These are very basic concepts and descriptions and are at the core of many chapters. We expect the readers to accept this explanation in a good sense.

B. K. Sinha et al., *Optimal Mixture Experiments*, Lecture Notes in Statistics 1028,
DOI: 10.1007/978-81-322-1786-2_3, © Springer India 2014

Let $\boldsymbol{x} = (x_1, x_2, \ldots, x_q)$ denote the vector of proportions of q mixing components and $\eta(\boldsymbol{x})$ be the corresponding mean response. Obviously, the factor space is the simplex, given by

$$\mathcal{X} = \{\boldsymbol{x} | x_i \geq 0, i = 1, 2, \ldots, q, \quad \Sigma x_i = 1\}. \tag{3.1.1}$$

Scheffé (1958) introduced the following models in *canonical* forms of different degrees to represent the mean response function $\eta(\boldsymbol{x})$:

$$\text{Linear: } \eta(\boldsymbol{x}) = \sum_i \beta_i x_i; \tag{3.1.2}$$

$$\text{Quadratic: } \eta(\boldsymbol{x}) = \sum_i \beta_i x_i + \sum_{i<j} \beta_{ij} x_i x_j; \tag{3.1.3}$$

$$\begin{aligned}\text{Full Cubic: } \eta(\boldsymbol{x}) = & \sum_i \beta_i x_i + \sum_{i<j} \beta_{ij} x_i x_j + \sum_{i<j<k} \beta_{ijk} x_i x_j x_k \\ & + \sum_{i<j} \delta_{ij} x_i x_j (x_i - x_j);\end{aligned} \tag{3.1.4}$$

$$\text{Special Cubic: } \eta(\boldsymbol{x}) = \sum_i \beta_i x_i + \sum_i \beta_i x_i + \sum_{i<j} \beta_{ij} x_i x_j + \sum_{i<j<k} \beta_{ijk} x_i x_j x_k. \tag{3.1.5}$$

Using the identity $\Sigma x_i = 1$, model (3.1.3) can be converted to the canonical *homogeneous* quadratic model:

$$\eta(\boldsymbol{x}) = \sum_i \beta_{ii} x_i^2 + \sum_{i<j} \beta_{ij} x_i x_j. \tag{3.1.6}$$

In the present study, we shall only be concerned with the canonical models (3.1.2) and (3.1.6). It is to be noted that the use of the identity $\Sigma x_i = 1$ may lead to remodeling very deceptively. For example, the quadratic model (3.1.3) may be modified to the special cubic model, thus inviting more parameters and more design points for estimation. Also proper interpretation of the parameters may be a bit complicated.

Scheffé (1958, 1963) introduced simplex lattice designs and simplex centroid designs for estimation of the parameters in the mean response function. Later on, considering only the interior points of the simplex, axial designs were introduced (cf. Cornell 2002). Optimality of mixture designs for the estimation of parameters was considered by Kiefer (1961, 1975), Atwood (1969), Galil and Kiefer (1977), Draper and Pukelsheim (1999), Pal et al. (2011), and Pal and Mandal (2012), among others. Following Sinha et al. (2010) we shall mainly discuss the estimability issues involving the parameters in the above mixture models. The optimality aspects will be taken up in subsequent chapters.

3.2 Some Standard Mixture Designs and Estimation of Parameters in Homogeneous Linear Mixture Models

There are some "standard" mixture designs suggested in the literature. These ensure, with reference to estimation of parameters in standard homogeneous models, as discussed above, the following:

1. full column rank condition;
2. general structure for the information matrix for routine computation of the variance and covariance of the estimates of the model parameters.

The following three are the most commonly used standard designs.

A. **Simplex Lattice Designs**: This class of designs consists of all feasible combinations of the mixing proportions wherein each proportion comprises of the values $(0, 1/m, 2/m, \ldots, m/m = 1)$ for a given integer parameter $m > 1$. Though there are $(m+1)^q$ possible combinations, only those combinations $(x_1, x_2, \ldots, x_q)$ are feasible which satisfy $x_1 + x_2 + \cdots + x_q = 1$. In a lattice design with q components and a given integer parameter m, the support set of design points is called a (q, m) simplex lattice. For a (q, m) simplex lattice design, there are $C(q+m-1, m)$ design points.

B. **Simplex Centroid Designs**: The centroid of a set of q nonzero coordinates in a q-dimensional coordinate system is the unique point $(1/q, 1/q, \ldots, 1/q)$. On the other hand, centroid of a set of t nonzero coordinates in a q-dimensional coordinate system is not unique. There are $C(q, t)$ centroid points of the form $(1/t, 1/t, \ldots, 1/t, 0, 0, \ldots, 0)$; $(1/t, 1/t, \ldots, 0, 1/t, 0, 0, \ldots, 0)$; ...... $(0, 0, \ldots, 0, 1/t, 1/t, \ldots, 1/t)$.

A simplex centroid design deals exclusively with the centroids of the coordinate systems, starting with exactly one nonzero component in the mixture [having q centroid points] and extending up to q nonzero components [having unique centroid point displayed above]. Thus, a simplex centroid design in the q-dimensional coordinate system contains $2^q - 1$ points.

As an example, for $q = 4$, there are 15 points altogether in the simplex centroid design : (1/4, 1/4, 1/4, 1/4); (1/3, 1/3, 1/3, 0); (1/3, 1/3, 0, 1/3); (1/3, 0, 1/3,1/3); (0, 1/3, 1/3, 1/3); (1/2, 1/2, 0, 0); (1/2, 0, 1/2, 0); (1/2, 0, 0,1/2); (0, 1/2, 1/2, 0); (0,1/2, 0, 1/2); (0, 0, 1/2, 1/2); (1, 0, 0, 0); (0, 1, 0, 0); (0, 0, 1, 0); (0, 0, 0, 1).

C. **Axial Designs**: It is to be noted that both simplex lattice and simplex centroid designs contain boundary points, i.e., points on the vertices, edges, and faces, *except* the centroid point $(1/q, 1/q, \ldots, 1/q)$, which lies inside the simplex. On the other hand, designs with interior points on the axis joining the points $x_i = 0, x_j = 1/(q-1)\ \forall j (\neq i)$ and $x_i = 1, x_j = 0\ \forall j (\neq i)$ are called axial designs. Thus, the axial design contains points of the form $[\{1 + (q-1)\Delta\}/q, (1-\Delta)/q, \ldots, (1-\Delta)/q]$, and its permutations, $-1/(q-1) < \Delta < 1$.

Let $\boldsymbol{x}_u = (x_{1u}, x_{2u}, \ldots, x_{qu})$, with $\sum_{i=1}^{q} x_{iu} = 1, u = 1, 2, \ldots, N$, be N design points. For any of the above standard designs, it may be noted that every ordered pair of proportions used in the design occurs equal number of times in any two columns of the design matrix. This implies that the information matrix is completely symmetric in the sense that all diagonal elements are equal to θ and all off-diagonal elements are equal to δ, i.e.,

$$\sum_{u=1}^{N} x_{iu}^2 = \theta \;\forall\, i,\; \sum_{u=1}^{N} x_{iu}x_{i'u} = \delta \;\forall\, i \neq i'. \tag{3.2.1}$$

Again the restrictions $\sum_{i=1}^{q} x_{iu} = 1, \quad u = 1, 2, \ldots, N$, imply that

$$\theta + (q-1)\delta = N/q. \tag{3.2.2}$$

So it is enough to find the explicit expression for either θ or δ. It also suggests that we can deduce algebraic expressions for both θ and δ and develop an identity involving design parameters.

We now give below expressions for the elements of the information matrices for the individual designs.

1. **Simplex Lattice Designs**: For (q, m) lattice design consisting of $C(q+m-1, m)$ points, it can be readily seen that

$$\theta = \sum_{r}(r/m)^2 C(m-r+q-2, m-r)] \tag{3.2.3}$$

$$\delta = \sum \sum_{r \leq t, r+t \leq m} (r/m)(t/m)[2-\delta_{(r,t)}]C(m-r-t+q-3, m-r-t) \tag{3.2.4}$$

where $\delta_{r,t}$ = Kronecker's delta.

It follows as a general rule that $\theta + (q-1)\delta = N/q$ holds good in all cases. Also the information matrix is positive definite, so that all the parameters are estimable.

2. **Simplex Centroid Designs**: In this design with q components, we have $N = 2^q - 1$.

Further, it is easy to verify that

$$\theta = \sum_{r>0}(1/r)^2 C(q-1, r-1) \tag{3.2.5}$$

$$\delta = \sum_{r>0}(1/r)^2 C(q-2, r-2). \tag{3.2.6}$$

It follows that, $\theta + (q-1)\delta = \sum_{r>0} C(q, r)/q = N/q$.

Here, also the information matrix is positive definite, so that all parameters are estimable.

3. **Axial Designs**:For given q [the number of mixing components], let "Δ" be a given quantity satisfying the condition: $-1/(q-1) < \Delta < 1$. Consider the $N = q$ design

points indexed by Δ:

$$[\{1 + (q - 1)\Delta\}/q, (1 - \Delta)/q, \ldots, (1 - \Delta)/q\} \text{ and its permutations}]. \quad (3.2.7)$$

It follows that for the mixture design exclusively based on the above set of q points,

$$\theta = [\{1 + (q - 1)\Delta\}^2 + (q - 1)(1 - \Delta)^2]/q^2 = \{1 + (q - 1)\Delta^2\}/q \quad (3.2.8)$$

and

$$\delta = [2(1 - \Delta)\{1 + (q - 1)\Delta\} + (q - 2)(1 - \Delta)^2]/q^2 = (1 - \Delta^2)/q. \quad (3.2.9)$$

It is easy to see that $\theta + (q - 1)\delta = 1 = N/q$, whatever be the choice of $\Delta, -1/(q - 1) < \Delta < 1$. Further, $\theta > \delta$, since $0 < \Delta^2 < 1$, and hence, the information matrix is positive definite so that all the parameters are estimable.

Remark 3.2.1 For homogeneous linear mixture models of the type (3.1.2), since the number of parameters is the same as the number of mixing components (q), the order of the information matrix is q and it has been straightforward to assert that the information matrices are positive definite for the standard mixture designs. This settles the question of estimability of the underlying linear model parameters. However, for homogeneous quadratic mixtures, asserting the positive definiteness of an information matrix is far from being a routine task, even for standard mixture designs. Therefore, it is advisable that one bypasses the problem of direct verification of positive definiteness of the information matrices and instead argues in a way to establish estimability of the parameters for standard mixture designs with a reasonable number of support points. This has been explicitly demonstrated in Sect. 3.4.

3.3 Generalizations of Axial Design and Their Comparison

3.3.1 Generalized Axial Design of Type I (D_1)

For a specified Δ, t copies of the axial design (3.2.7) are taken. We call the generated design as Generalized Axial Design of Type I. Obviously, $N = qt$ and the information matrix of this design is completely symmetric with diagonal elements as θ_1 and off-diagonal elements as δ_1, where

$$\theta_1 = t\{1 + (q - 1)\Delta^2\}/q \quad (3.3.1)$$

and

$$\delta_1 = t(1 - \Delta^2)/q. \quad (3.3.2)$$

3.3.2 Generalized Axial Design of Type II (D_2)

The design is obtained by union of t axial designs of the type (3.2.7) with the parameters $\Delta_1, \Delta_2, \ldots, \Delta_t, -1/(q-1) < \Delta_i < 1, i = 1, 2, \ldots, t$. It is implicitly understood that not all the Δs are equal. Obviously, $N = qt$ and the information matrix is completely symmetric with diagonal elements as θ_2 and off-diagonal elements as δ_2, where

$$\theta_2 = \sum_i [1 + (q-1)\Delta_i^2]/q \tag{3.3.3}$$

$$\delta_2 = \sum_i (1 - \Delta_i^2)/q. \tag{3.3.4}$$

We want to compare the two designs in terms of their information matrices both of which are completely symmetric. The difference in the information matrices of the two designs is also completely symmetric with diagonal elements as

$$\theta_1 - \theta_2 = (q-1)[t\Delta^2 - \Sigma_i \Delta_i^2]/q = a, \text{ say.} \tag{3.3.5}$$

Also the difference matrix has row [column] sums zero. Then, it follows that the off-diagonal elements of the difference matrix are all equal to $[-a/(q-1)]$. Hence, the difference matrix is nonnegative definite iff $a > 0$. In order to make a comparison between the two designs, we now assume that $t\Delta = \Sigma_i \Delta_i$. Subject to this restriction, it turns out that $a < 0$. Hence, we get the following theorem.

Theorem 3.3.1 *Under the condition $\Delta = \frac{1}{t}\Sigma_{i=1}^{q} \Delta_i$, Generalized Axial Design of Type II dominates Generalized Axial design of Type I in terms of their information matrices.*

3.3.3 Generalized Axial Design of Type III (D_3)

Let $f_0, f_1, \ldots, f_p$ be $p+1$ proper positive integers with $\Sigma_{i=0}^{p} f_i = q$. Then, the design with $N = q!/\Pi\, f_i!$ points, given by

$$\left\{1 + \sum_{i=1}^{p} f_i\Delta_i)/q, \ldots, (1 + \sum_{i=1}^{p} f_i\Delta_i)/q, (1-\Delta_1)/q, \ldots, (1-\Delta_1)/q, \ldots,\right.$$
$$\left.(1-\Delta_p)/q, \ldots, (1-\Delta_p)/q\right\} \tag{3.3.6}$$

and their permutations, where $(1 + \Sigma_{i=1}^{p} f_i \Delta_i)/q$ is repeated f_0 times and $(1-\Delta_i)/q$ is repeated f_i times, and $i = 1, 2, \ldots, p$, is called a Generalized Axial Design of Type III with parameters as stated above. This design is denoted as

$$D_3 = \left\{ \left(\frac{1 + \sum_{i=1}^{p} f_i \Delta_i}{q} \right)^{f_0}, \left(\frac{1 - \Delta_i}{q} \right)^{f_i}, \quad i = 1, 2, \ldots, p \right\}. \tag{3.3.7}$$

Note that Δ_i's must satisfy

$$0 < \min\{1 + \sum_i f_i \Delta_i\}/q; \left(\frac{1 - \Delta_i}{q} \right), \quad i = 1, 2, \ldots, p\}$$
$$< \max\{1 + \sum_i f_i \Delta_i\}/q; \left(\frac{1 - \Delta_i}{q} \right), \quad i = 1, 2, \ldots, p\} < 1. \tag{3.3.8}$$

The permutations jointly give rise to N design points and also lead to complete symmetry of the underlying information matrix.

Example 3.3.1 Consider $q = 4$, $p = 2$; $f_0 = 1$, $f_1 = 1$, $f_2 = 2$. Explicitly written, with this combination, there are $N = 12$ design points as displayed below:

$$[(1 + \Delta_1 + 2\Delta_2]/4, (1 - \Delta_1)/4, (1 - \Delta_2)/4, (1 - \Delta_2)/4];$$
$$[(1 + \Delta_1 + 2\Delta_2]/4, (1 - \Delta_2)/4, (1 - \Delta_1)/4, (1 - \Delta_2)/4];$$
$$[(1 + \Delta_1 + 2\Delta_2]/4, (1 - \Delta_2)/4, (1 - \Delta_2)/4, (1 - \Delta_1)/4];$$
$$[(1 - \Delta_1)/4, (1 + \Delta_1 + 2\Delta_2]/4, (1 - \Delta_2)/4, (1 - \Delta_2)/4];$$
$$[(1 - \Delta_2)/4, (1 + \Delta_1 + 2\Delta_2]/4, (1 - \Delta_1)/4], (1 - \Delta_2)/4];$$
$$[(1 - \Delta_2)/4, (1 + \Delta_1 + 2\Delta_2]/4], (1 - \Delta_2)/4, (1 - \Delta_1)/4];$$
$$[(1 - \Delta_1)/4, (1 - \Delta_2)/4, (1 + \Delta_1 + 2\Delta_2]/4, (1 - \Delta_2)/4];$$
$$[(1 - \Delta_2)/4, (1 - \Delta_1)/4, (1 + \Delta_1 + 2\Delta_2]/4, (1 - \Delta_2)/4];$$
$$[(1 - \Delta_2)/4, (1 - \Delta_2)/4, (1 + \Delta_1 + 2\Delta_2]/4, (1 - \Delta_1)/4];$$
$$[(1 - \Delta_1)/4, (1 - \Delta_2)/4, (1 - \Delta_2)/4, (1 + \Delta_1 + 2\Delta_2]/4];$$
$$[(1 - \Delta_2)/4, (1 - \Delta_1)/4, (1 - \Delta_2)/4, (1 + \Delta_1 + 2\Delta_2]/4];$$
$$[(1 - \Delta_2)/4, (1 - \Delta_2)/4, (1 - \Delta_1)/4, (1 + \Delta_1 + 2\Delta_2]/4].$$

In the following discussions, we assume

$$f_0 = 1, x_0 = [1 + \sum_{i=1}^{p} f_i \Delta_i]/q, x_i = (1 - \Delta_i)/q \text{ for } i = 1, 2, \ldots, p. \tag{3.3.9}$$

In terms of the x's, it follows that the common diagonal element θ_3 and off-diagonal element δ_3 are given by

$$\theta_3 = x_0^2 N/q + \Sigma x_i^2 f_i [N/q] \tag{3.3.10}$$

$$\begin{aligned}\delta_3 =& 2x_0 \sum_{i=1}^{p} f_i [N/q(q-1)] + \sum_{i=1}^{p} x_i^2 f_i (f_i - 1)[N/q(q-1)] \\ &+ 2 \sum_{i<j} x_i x_j f_i f_j [N/q(q-1)].\end{aligned} \tag{3.3.11}$$

Note that the δs do not depend on the particular pair of components in the product moments.

Again the identity:

$$\left(1 + \sum_i f_i \Delta_i\right) + \sum_i f_i (1 - \Delta_i) = q \tag{3.3.12}$$

implies, as expected, that

$$\theta_3 + (q-1)\delta_3 = N/q.$$

Using (3.3.10) and (3.3.11), it can be proved that

$$\theta_3 - \delta_3 = [N/(q-1)][\Sigma(x_i - \overline{x})^2 f_i] \geq 0, \tag{3.3.13}$$

where

$$\overline{x} = \sum_{i=0}^{p} f_i x_i / \sum_i f_i = 1/q. \tag{3.3.14}$$

Since Δ_is in (3.3.9) are unequal, we have $\theta_3 > \delta_3$.

Therefore, the information matrix is positive definite, whatever be the choice of the Δ_is, all distinct from one another, for a given $p \geq 2$.

3.3.4 Comparison Between Generalized Axial Designs of Type II (D_2) and Type III (D_3)

Let D_2 be written as

$$D_2 = \bigcup_{i=1}^{t} D_{2i}, \tag{3.3.15}$$

where

$$D_{2i} = \left(\frac{1 + (q-1)\Delta_i'}{q}, \frac{1 - \Delta_i'}{q}, \ldots, \frac{1 - \Delta_i'}{q} \text{ and their permutiations}\right) \tag{3.3.16}$$

D_3 is given by (3.3.7). We assume $f_0 = 1$. To compare D_2 with D_3, we make the following reasonable assumptions:

a. Number of observations in D_2 and D_3 is the same, i.e.,

$$N^* = q!/f_i! = qt, \tag{3.3.17}$$

implying

$$t = N^*/q = (1/q)q!/f_i!. \tag{3.3.18}$$

b. The average of Δ_js of D_2 is the same as the average of Δ_is of D_3, i.e.,

$$\overline{\Delta} = \frac{1}{t}\sum_{j=1}^{t} \Delta_j' = \frac{1}{q-1}\sum_{i=1}^{p} f_i \Delta_i. \tag{3.3.19}$$

From (3.3.3), (3.3.10), (3.3.17), and (3.3.19), the common diagonal element of the difference in the information matrices of D_2 and D_3 can be found to be

$$\theta_2 - \theta_3 = \frac{(q-1)N^*}{q^3}[\sigma_\Delta^2 - q\sigma_{\Delta'}^2], \tag{3.3.20}$$

where

$$\sigma_\Delta^2 = \frac{1}{q-1}\sum_{i=1}^{p} f_i(\Delta_i - \overline{\Delta})^2 \tag{3.3.21}$$

$$\sigma_{\Delta'}^2 = \frac{1}{t}\sum_{j=1}^{t} (\Delta_j' - \overline{\Delta})^2. \tag{3.3.22}$$

So, by the same reasoning as given in proving Theorem 3.3.1, it follows that D_2 dominates D_3 iff $\theta_2 - \theta_3 > 0$, i.e., iff $\sigma_\Delta^2 > k\sigma_{\Delta'}^2$. So we get the following theorem:

Theorem 3.3.2 *Under the conditions* (3.3.18) *and* (3.3.19), *Generalized Axial design of Type II dominates Generalized Axial design of Type III iff* $\sigma_\Delta^2 > k\sigma_{\Delta'}^2$, *where* σ_Δ^2 *and* $\sigma_{\Delta'}^2$ *are given by* (3.3.21) *and* (3.3.22), *respectively.*

3.4 Estimation of Parameters in Canonical Homogeneous Quadratic Mixture Model

In the homogeneous quadratic mixture model (3.1.6), in its canonical form, there are $C(q+1, 2)$ parameters. It would be interesting to address the [unbiased] estimation issue for all the parameters on the basis of the designs introduced above.

Let, as before, $x_u = (x_{1u}, x_{2u}, \ldots, x_{qu})$, $\sum_{i=1}^{q} x_{iu} = 1 \ \forall\ u = 1, 2, \ldots, N$ be the N experimental points. The uth row of the design matrix is $\left(x_{1u}^2, x_{2u}^2, \ldots, x_{qu}^2, x_{1u}x_{2u}, x_{1u}x_{3u}, \ldots, x_{q-1u}x_{qu}\right)$, $u = 1, 2, \ldots, N$.

A typical element of the information matrix is $\left[\alpha_1, \alpha_2, \ldots, \alpha_q\right] = \sum_u x_{1u}^{\alpha_1} x_{2u}^{\alpha_2} \cdots x_{qu}^{\alpha_q}$, where $\alpha_i \geq 0$, $\sum_i \alpha_i = 4$.

Unlike in the case of homogeneous linear mixture models, it turns out that the information matrix for a homogeneous quadratic mixture model is not completely symmetric in the sense of the diagonal elements being equal to each other and also the off-diagonal elements being equal to each other.

However, the information matrix is symmetric in a generalized sense. This is the sense of invariance with respect to all possible permutations of the components. This is due to the fact that in all the above designs, the proportions of any experimental point are permutated among the components. Five distinct elements denoted by a, b, etc., of the information matrix correspond to

(i) $[4] = \sum_u x_{(i,u)}^4 = a,$

(ii) $[2, 2] = \sum_u x_{(i,u)}^2 x_{(j,u)}^2 = b,$

(iii) $[3, 1] = \sum_u x_{(i,u)}^3 x_{(j,u)} = c,$

(iv) $[2, 1, 1] = \sum_u x_{(i,u)}^2 x_{(j,u)} x_{(r,u)} = d,$

(v) $[1, 1, 1, 1] = \sum_u x_{(i,u)} x_{(j,u)} x_{(r,u)} x_{(s,u)} = e.$ (3.4.1)

In the above, it is to be understood that we are referring to all possible $1 \leq i \neq j \neq r \neq s \leq q$.

The information matrix is as such a block matrix having the composition

$$I = \begin{matrix} \beta_{ii}s & \beta_{ij}s \\ \end{matrix} \atop \begin{pmatrix} A & B \\ & C \end{pmatrix}, \qquad (3.4.2)$$

where

$A = ((a, b, b, \ldots, b))$ is a circulant [square] matrix of order q,
$B = ((\ldots ccccccc \ldots dddd))$ is a matrix of order $q \times C(q, 2)$ with each row containing $(q-1)$ cs, possibly forming a single run, and ds for all other entries,
$C = ((\ldots dddd \ldots eeee \ldots b \ldots eeee \ldots dddd))$ is a square matrix of order $C(q, 2)$ where in each row b appears only once in the diagonal position, d appears in $2(q-2)$ positions, and e appears in all other positions.

Next, we study the designs introduced above for estimation of the parameters in the model (3.1.6) and find the elements of the information matrices. As stated in Remark 3.2.1, we proceed as follows.

A. **Estimability of parameters using (q, m) simplex lattice design $(m \geq 2)$**

As described in Sect. 3.2 a (q, m) lattice design contains the q extreme points which are permutations of $(1, 0, \ldots, 0)$ and the set of $C(q, 2)$ points which are permutations of $\left(\frac{1}{m}, \frac{m-1}{m}, 0, \ldots, 0\right)$ *among* other points. These are sufficient for unbiased estimation of β_{ii}s and β_{ij}s involved in the model (3.1.6).

Information Matrix

Here, the expressions for the quantities in (i) to (v) in (3.4.1) are as follows:

$$a = \sum_{r=1}^{m} \left(\frac{r}{m}\right)^4 C(m - r + q - 2, m - r)$$

$$b = \sum_{\substack{1 \leq r < t \leq m \\ r+t \leq m}} \left(\frac{r}{m}\right)^2 \left(\frac{t}{m}\right)^2 (2 - \delta_{rt}) C(m - r - t + q - 3, m - r - t)$$

$$c = \sum_{\substack{1 \leq r < t \leq m \\ r+t \leq m}} \left(\frac{r}{m}\right)^3 \left(\frac{t}{m}\right) (2 - \delta_{rt}) C(m - r - t + q - 3, m - r - t)$$

$$\begin{aligned} d = & 6 \sum_{\substack{1 \leq r < t \leq m \\ r+t \leq m}} \left(\frac{r}{m}\right)^2 \left(\frac{t}{m}\right) \left(\frac{s}{m}\right) C(m - r - t - s + q - 4, m - r - t - s) \\ & + 3 \sum_{\substack{1 \leq r < t \leq m \\ r+t \leq m}} \left(\frac{r}{m}\right)^3 \left(\frac{t}{m}\right) (2 - \delta_{rt}) C(m - 2r - t + q - 4, m - 2r - t) \\ & + \sum_{\substack{1 \leq r < t \leq m \\ r+t \leq m}}^{m} \left(\frac{r}{m}\right)^4 C(m - 3r + q - 4, m - 3r) \end{aligned}$$

$$\begin{aligned} e = & 24 \sum_{\substack{1 \leq r < t \leq m \\ r+t+s+p \leq m}} \left(\frac{r}{m}\right) \left(\frac{t}{m}\right) \left(\frac{s}{m}\right) \left(\frac{p}{m}\right) \\ & \times C(m - r - t - s - p + q - 4, m - r - t - s - p) \\ & + 12 \sum_{\substack{1 \leq r < t \leq m \\ 2r+t+s \leq m}} \left(\frac{r}{m}\right)^2 \left(\frac{t}{m}\right) \left(\frac{s}{m}\right) \\ & \times C(m - 2r - t - s + q - 5, m - 2r - t - s) \\ & + 4 \sum_{\substack{1 \leq r < t \leq m \\ 3r+t \leq m}}^{m} \left(\frac{r}{m}\right)^3 \left(\frac{t}{m}\right) C(m - 3r - t + q - 4, m - 3r - t) \\ & + \sum_{\substack{1 \leq r < t \leq m \\ 4r \leq m}} \left(\frac{r}{m}\right)^4 C(m - 4r + q - 5, m - 4). \end{aligned} \quad (3.4.3)$$

B. **Estimability using simplex centroid designs**

Among the $N(q) = 2^q - 1$ points of the centroid design, there are points which are permutations of $(1, 0, \ldots, 0)$ and $(1/2, 1/2, 0, \ldots, 0)$. These $C(q+1, 2)$ points are sufficient to estimate the $C(q+1, 2)$ parameters of quadratic model (3.1.6) unbiasedly.

Information Matrix
The expressions for the quantities (i) to (v) in (3.4.1) are given below:

$$\begin{aligned} a &= \sum_{r=1}^{q} C(q-1, r-1)\left(\frac{1}{r}\right)^4 \\ b &= \sum_{r=2}^{q} C(q-2, r-2)\left(\frac{1}{r}\right)^4 = c \\ d &= \sum_{r=3}^{q} C(q-3, r-3)\left(\frac{1}{r}\right)^4 \\ e &= \sum_{r=4}^{q} C(q-4, r-4)\left(\frac{1}{r}\right)^4 . \end{aligned} \tag{3.4.4}$$

C. **Estimability using axial-type designs**

The standard axial design (3.2.7) contains only q points, and as such, these points cannot estimate $C(q+1, 2)$ parameters of the quadratic model (3.1.6). For the case of homogeneous quadratic mixture model, we assert below that a set of $N = 2q + C(q, 2)$ design points ensures estimability of all the $[q + C(q, 2)]$ parameters.

For this, we define the following sets of design points:

$$\begin{aligned} S_{\Delta_1} = \text{set of all permutations of } [\{1 + (q-1)\Delta_1\}/q, \\ (1-\Delta_1)/q, \ldots, (1-\Delta_1)/q], \end{aligned} \tag{3.4.5}$$

$$\begin{aligned} S_{\Delta_2} = \text{set of all permutations of } [\{1 + (q-1)\Delta_2\}/q, \\ (1-\Delta_2)/q, \ldots, (1-\Delta_2)/q], \end{aligned} \tag{3.4.6}$$

taking $0 < \Delta_1 \neq \Delta_2 < 1$.

Note that $S_{\Delta_1} \cup S_{\Delta_2}$ gives a Generalized Axial Design of Type II with $t = 2$. Additionally, we define

$$\begin{aligned} S_{\Delta} = \text{set of all permutations of } [(1 + \{(q-2)\Delta/2\})/q, \\ (1 + \{(q-2)\Delta/2\})/q; \{1-\Delta\}/q, \ldots, \{1-\Delta\}/q] \end{aligned} \tag{3.4.7}$$

which gives a Generalized Axial Design of Type III with $f_0 = 2, f_1 = f_2 = \ldots f_{q-2} = 1$.

We consider the set $S = S_{\Delta_1} \cup S_{\Delta_2} \cup S_{\Delta}$ of $2q + C(q, 2)$ design points which may be termed as Generalized Axial Design of Type IV (D_4).

Let $y_{i1}, y_{i2}, \ldots, y_{iq}$ be the observations corresponding to the design points in $S_{\Delta_i}, i = 1, 2$. Then,

$$E[y_{11}] = \beta_{11} P_1^2 + Q_1^2[\beta - \beta_{11}] + P_1 Q_1[\underset{\sim}{\beta}_1] + Q_1^2[\gamma - \underset{\sim}{\beta}_1] \tag{3.4.8}$$

where

$$\beta = \beta_{11} + \beta_{22} + \cdots + \beta_{qq}$$
$$\underset{\sim}{\beta}_1 = \beta_{12} + \beta_{13} + \cdots + \beta_{1q}$$
$$\gamma = \sum_{i=1}^{q} \sum_{j=i+1}^{q} \beta_{ij} \tag{3.4.9}$$
$$P_1 = \{1 + (q-1)\Delta_1\}/q \text{ and } Q_1 = (1 - \Delta_1)/q. \tag{3.4.10}$$

Likewise, $E(y_{ij})$ can be computed for $i = 1, 2$; $j = 1, 2, ..., q$. Adding the q expectations $E(y_{1i}), i = 1, 2, \ldots, q$, the expectation of the total response y_{10} comes out to be

$$E(y_{10}) = [P_1^2 + (q-1)Q_1^2]\beta + [2P_1Q_1 + (q-2)Q_1^2]\gamma. \tag{3.4.11}$$

Similarly, from the points in (3.4.6), and using the obvious notations for the resulting observations, it may be derived that

$$E(y_{20}) = [P_2^2 + (q-1)Q_2^2]\beta + [2P_2Q_2 + (q-2)Q_2^2]\gamma, \tag{3.4.12}$$

where $P_2 = (1 + (q-1)\Delta_2)/q$, $Q_2 = (1 - \Delta_2)/q$.

From (3.4.11) and (3.4.12), we can solve for β and γ. The determinant of the 2×2 coefficient matrix turns out to be

$$(P_1Q_2 - P_2Q_1)[P_2(2 - qQ_1) - Q_2(2 - qP_1)]. \tag{3.4.13}$$

Our choice of the design parameters Δ_1 and Δ_2 should ensure that the above is nonzero. Next, we make use of (3.4.9) and its analog from S_{Δ_2} [in terms of P_2 and Q_2] to solve for β_{11} and $\underset{\sim}{\beta}_1$. This is possible whenever the underlying 2×2 coefficient matrix is non-singular. This happens whenever

$$(P_1Q_2 - P_2Q_1)(P_1 - Q_1)(P_2 - Q_2) \neq 0. \tag{3.4.14}$$

Note that our choice of $\Delta_1 > 0$ and $\Delta_2 > 0$ ensures $P_1 > Q_1$ and $P_2 > Q_2$.

Therefore, in effect, (3.4.15) requires $(P_1Q_2 - P_2Q_1) \neq 0$, i.e., the first factor in (3.4.14) to be nonzero. Hence, (3.4.15) is a built-in condition in (3.4.14). This is how all the β_{ii}s and $\underset{\sim}{\beta}_i$s are estimated.

We have yet to ascertain estimability of the β_{ij}s for $i \neq j$. This time, we will use the points of (3.4.7).
Set

$$P = \{1 + (q-2)\Delta/2\}/q \text{ and } Q = (1-\Delta)/q \qquad (3.4.15)$$

so that $2P + (q-2)Q = 1$. The model expectation of the observation corresponding to the design point $[P, P, Q, Q, \ldots, Q]$ in (3.4.7) involves the only unknown parameter β_{12}, apart from the parameters $\beta, \beta_{11}, \beta_{22}, \gamma, \tilde{\beta}_1$, and $\tilde{\beta}_2$, which are already estimated. It turns out that the coefficient of β_{12} is given by $(P-Q)^2 + Q^2$ which is positive.

Thus, we get the following theorem.

Theorem 3.4.3 *The Generalized Axial Design of Type IV ensures estimability of the parameters of canonical homogeneous quadratic model of* (3.1.6).

Information Matrix
The expressions of the quantities (i) to (v) of (3.4.1), which are the elements of the information matrix of the design, are given below:

$$\begin{aligned}
a &= (p_1^4 + p_2^4) + (q_1^4 + q_2^4)(q-1) + (q-1)\left(P^4 + Q^4\frac{(q-2)}{2}\right)\\
b &= P^4 + 2[p_1^2q_1^2 + p_2^2q_2^2 + P^2Q^2(q-2)] + Q^4C(q-2,2)\\
&\quad + (q_1^4 + q_2^4)(q-2)\\
c &= p_1^3q_1 + q_1^3p_1 + q_1^4 + p_2^3q_2 + q_2^3p_2 + P^4 + (P^3Q + Q^3P)(q-2)\\
&\quad + (q_1^4 + q_2^4)(q-2) + Q^4C(q-2,2)\\
d &= 2P^3Q + (q-2)P^2Q^2 + 2(q-3)Q^3P + Q^4C(q-3,2)\\
&\quad + [p_1^2q_1^2 + 2q_1^3p_1 + (q-3)q_1^4]\\
e &= 4(p_1q_1^3 + p_2q_2^3) + 6P^2Q^2 + 4(q-4)PQ^3 + C(q-4,2)Q^4
\end{aligned}$$

3.5 Other Mixture Models

As was mentioned in Chap. 1, in the literature, some other mixture models have been introduced and studied. Below we take up symmetrized version of two such models, viz., Becker's homogeneous model of degree one [vide (1.2.7)] and Draper–St. John's model [vide (1.2.11)] and discuss the estimability issues at length.

3.5.1 Estimation of Parameters in Becker's Homogeneous Model of Degree One in the Presence of Only Two-Component Synergistic Effects

The model (1.2.7) is rewritten below:

$$E(y) = \beta_1 x_1 + \beta_2 x_2 + \cdots + \beta_q x_q + \beta_{12}\frac{x_1 x_2}{x_1 + x_2} + \cdots + \beta_{q-1q}\frac{x_{q-1}x_q}{x_{q-1} + x_q}$$
$$0 \le x_i \le 1, \forall i, x_i + x_j > 0 \ \forall\ i \ne j. \tag{3.5.1}$$

Below we describe a design D based on $2q + C(q, 2)$ points which ensure estimability of all the model parameters. It is obtained as union of three subdesigns, i.e., $D = \overset{3}{\underset{i=1}{U}} D_i$, where

$$\underset{\text{Subdesign 1 }(D_1)}{\begin{pmatrix} b_1 & a_1 & a_1 & \ldots & a_1 \\ a_1 & b_1 & a_1 & \ldots & a_1 \\ \vdots & \vdots & \vdots & \vdots & \vdots \\ a_1 & a_1 & a_1 & \ldots & b_1 \end{pmatrix}} \quad \underset{\text{Subdesign 2 }(D_2)}{\begin{pmatrix} b_2 & a_2 & a_2 & \ldots & a_2 \\ a_2 & b_2 & a_2 & \ldots & a_2 \\ \vdots & \vdots & \vdots & \vdots & \vdots \\ a_2 & a_2 & a_2 & \ldots & b_2 \end{pmatrix}} \quad \underset{\text{Subdesign }(D_3)}{\begin{pmatrix} b, b, a, a, \ldots, a \\ \text{and all } \binom{q}{2} \\ \text{permutations} \end{pmatrix}} \tag{3.5.2}$$

It is to be noted that D is actually a Generalized Axial Design of Type IV (D_4) described in Sect. 3.4C.
Define

$$\beta = \Sigma\beta_i, \gamma = \sum_{1 \le i < j \le q} \beta_{ij}, \tilde{\beta}_i = \beta_{i1} + \cdots + \beta_{i,i-1} + \beta_{i,i+1} + \cdots + \beta_{iq}; \quad i = 1, 2, \ldots, q.$$

Let y_{1i}, $y_{2i'}$ and $y_{3i''}$ denote ith, i'th and i''th observations of D_1, D_2, and D_3, respectively, $i, i' = 1, 2, \ldots, q, i'' = 1, 2, \ldots, C(q, 2)$.

$$E(y_{1i}) = b_1\beta_i + a_1(\beta - \beta_i) + \frac{a_1 b_1}{a_1 + b_1}\tilde{\beta}_i + \frac{a_1^2}{2a_1}(\gamma - \tilde{\beta}_i); \quad i = 1, 2, \ldots, q. \tag{3.5.3}$$

Summing over all i:

$$E(y_{10}) = [b_1 + (q-1)a_1]\beta + \left[\frac{2a_1 b_1}{a_1 + b_1} + (q-2)\frac{a_1}{2}\right]\gamma. \tag{3.5.4}$$

Similarly, from D_2, we have

$$E(y_{20}) = [b_2 + (q-1)a_2]\beta + \left[\frac{2a_2 b_2}{a_2 + b_2} + (q-2)\frac{a_2}{2}\right]\gamma. \tag{3.5.5}$$

From (3.5.4) and (3.5.5), we can solve for β and γ.
Consider the equations

$$E(y_{1i}) = (b_1 - a_1)\beta_i + a_1\beta + \frac{a_1}{2}\gamma + \left(\frac{a_1 b_1}{a_1 + b_1} - \frac{a_1}{2}\right)\underset{\sim}{\beta}_i \tag{3.5.6}$$

and

$$E(y_{2i}) = (b_2 - a_2)\beta_i + a_2\beta + \frac{a_2}{2}\gamma + \left(\frac{a_2 b_2}{a_2 + b_2} - \frac{a_2}{2}\right)\underset{\sim}{\beta}_i. \tag{3.5.7}$$

Substituting $\hat{\beta}$ and $\hat{\gamma}$ in (3.5.6) and (3.5.7), we can estimate β_i and $\underset{\sim}{\beta}_i, i = 1, 2, \ldots, q$. Though the part totals $\underset{\sim}{\beta}_i$s are estimated, β_{ij}s are still to be estimated individually. For this, we consider the points of D_3. Let us consider the design point $(b, b, a, a, \ldots, a)$ and the corresponding observation y_{31}. It can be derived that

$$\begin{aligned} E(y_{31}) = {} & (b - a)(\beta_1 + \beta_2) + a\beta + \frac{a}{2}\gamma + \left(\frac{ab}{a+b} - \frac{a}{2}\right)(\underset{\sim}{\beta}_1 + \underset{\sim}{\beta}_2) \\ & + \left(\frac{a+b}{2} - \frac{2ab}{a+b}\right)\beta_{12}. \end{aligned} \tag{3.5.8}$$

In (3.5.8), estimates of all parameters except β_{12} are known. So β_{12} can be estimated. From the $C(q, 2)$ observations of D_3, $C(q, 2)$ parameters, viz., $\beta_{12}, \beta_{13}, \ldots, \beta_{q-1,q}$ can be estimated in a similar fashion.

Thus, we get the following theorem.

Theorem 3.5.4 *Axial Design of Type IV ensures estimability of the parameters of the Becker's model* (3.5.1).

3.5.2 Another Competing Design

Below we provide yet another competing design based on $q(q - 1) + 1$ design points which also ensures estimability of all the model parameters in (3.5.1). This design is also easy to analyze, and this compares favorably with the earlier one.

3.5.3 Design Description and Estimability

The design D^* consisting of $N(q) = q(q - 1) + 1$ design points is defined as

$$D^* = D_1^* \cup D_0^*, \tag{3.5.9}$$

where

$$D_1^* = \left\{ \text{all permutations of } \left(0, \frac{2}{q}, \frac{1}{q}, \ldots, \frac{1}{q}\right)\right\}, D_0^* = \left(\frac{1}{q}, \frac{1}{q}, \frac{1}{q}, \ldots, \frac{1}{q}\right). \tag{3.5.10}$$

Let T = total of all observations on the points of D_1^*.
It can be easily shown that

$$E(T) = (q-1)\beta + \frac{(q-2)(3q-1)}{6q}\gamma, \tag{3.5.11}$$

where

$$\beta = \sum_{i=1}^{q} \beta_i, \gamma = \sum_{1 \le < j \le q} \beta_{ij}. \tag{3.5.12}$$

Again, let y be the observation at the experimental point of D_0^*.
Then,

$$E(y) = \frac{1}{q}\beta + \frac{1}{2q}\gamma. \tag{3.5.13}$$

From (3.5.11) and (3.5.13), β and γ can be estimated, as the determinant of the coefficient matrix of β and γ is $\frac{2q-1}{3q^2}$, which is positive for all values of q.

Now, we try to estimate

$$\tilde{\beta}_i = \sum_{\substack{j=1 \\ j \neq i}}^{q} \beta_{ij}; \quad i = 1, 2, \ldots, q. \tag{3.5.14}$$

Let us consider the following set of $2(q-1)$ design points from D_1^* and the corresponding observations

(3.5.15)

Design Points					Observations
x_1	x_2	x_3	$\ldots$	x_q	
0	$2/q$	$1/q$	$\cdots$	$1/q$	y_{12}
0	$1/q$	$2/q$	$\cdots$	$1/q$	y_{13}
$\vdots$	$\vdots$	$\vdots$	$\vdots$	$\vdots$	$\vdots$
0	$1/q$	$1/q$	$\cdots$	$2/q$	y_{1q}
$2/q$	0	$1/q$	$\cdots$	$1/q$	y_{22}
$2/q$	$1/q$	0	$\cdots$	$1/q$	y_{23}
$\vdots$	$\vdots$	$\vdots$	$\vdots$	$\vdots$	$\vdots$
$2/q$	$1/q$	$1/q$	$\cdots$	0	y_{2q}

It can be proved in the usual manner that

$$E\left\{\sum_{i=2}^{q}(y_{1i}+y_{2i})\right\}=\frac{2(q-1)}{q}\beta+\frac{(3q-5)}{3q}\gamma-\frac{(q-1)}{3q}\underset{\sim}{\beta}_1. \quad (3.5.16)$$

As the estimators of β and γ are already available, the estimator of $\underset{\sim}{\beta}_1$ can be obtained from (3.5.16).

Considering the sums of suitable pairs of observations, the estimators of other $\underset{\sim}{\beta}$s can be obtained. We have still to find the estimators of βs. For this, we consider the expectation of $\Sigma_{i=2}^{q}y_{1i}$, which is given by

$$E\left(\sum_{i=2}^{q}y_{1i}\right)=\beta+\frac{3q-1}{6q}\gamma-\frac{3q-1}{6q}\underset{\sim}{\beta}_1-\beta_1. \quad (3.5.17)$$

From Eq. (3.5.17), an estimator of β_1 can be obtained by using the estimators of β, γ, and $\underset{\sim}{\beta}_1$. Considering such suitable partial sums of y's, other β's can be estimated.

Below we provide a table indicating a comparison of the number of support points of the two competing designs.

q	Number of model Parameters	No. of design points	
		Design D	Design D^*
2	3	5	3
3	6	9	7
4	10	14	13
5	15	20	21
$q>5$	$C(q+1,2)$	$2q+C(q,2)$	$q(q-1)+1$

Remark 3.5.2 For $q\geq 5$, the number of design points in D is less than that in D^*.

3.5.4 Estimation of Parameters in Draper–St. John's Model

The model (1.2.11) is reproduced below:

$$\eta_x=\beta_1x_1+\cdots+\beta_qx_q+\frac{\alpha_1}{x_1}+\cdots+\frac{\alpha_q}{x_q},\quad 0<x_i<1\ \forall\ i. \quad (3.5.18)$$

Define $\beta=\Sigma\beta_i,\ \alpha=\Sigma\alpha_i$.

Consider the design D^{**} which is the union of the following two subdesigns D_1^* and D_2^*.

$$D_1^* = \overset{\text{Sub-design I}}{\begin{pmatrix} b_1 & a_1 & a_1 & \dots & a_1 \\ a_1 & b_1 & a_1 & \dots & a_1 \\ \vdots & \vdots & \vdots & \vdots & \vdots \\ a_1 & a_1 & a_1 & \dots & b_1 \end{pmatrix}} \quad D_2^* = \overset{\text{Sub-design II}}{\begin{pmatrix} b_2 & a_2 & a_2 & \dots & a_2 \\ a_2 & b_2 & a_2 & \dots & a_2 \\ \vdots & \vdots & \vdots & \vdots & \vdots \\ a_2 & a_2 & a_2 & \dots & b_2 \end{pmatrix}}$$

D^{**} is a generalized axial design of type II (D_2). Below we discuss the question of estimability of model parameters.

Let y_{1i} and y_{2j} be the ith and jth observations of D_1^* and D_2^*, respectively; $i, j = 1, 2, ..., q$. Then, it follows that

$$E(y_{1i}) = a_1\beta + \frac{\alpha}{a_1} + (b_1 - a_1)\beta_i + \left(\frac{1}{b_1} - \frac{1}{a_1}\right)\alpha_i; \quad i = 1, 2, \dots, q.$$

Adding over i, we get

$$E(y_{10}) = (b_1 + (q-1)a_1)\beta + \frac{(a_1 + (q-1)b_1)}{a_1 b_1}\alpha. \tag{3.5.19}$$

Similarly,

$$E(y_{20}) = (b_2 + (q-1)a_2)\beta + \frac{(a_2 + (q-1)b_2)}{a_2 b_2}\alpha. \tag{3.5.20}$$

From (3.5.19) and (3.5.20), we can estimate α and β.
Again from

$$E(y_{1i}) = \left(\alpha_1\beta + \frac{\alpha}{a_1}\right) + (b_1 - a_1)\beta_i + \frac{a_1 - b_1}{a_1 b_1}\alpha_i$$
$$E(y_{2i}) = \left(\alpha_2\beta + \frac{\alpha}{a_2}\right) + (b_2 - a_2)\beta_i + \frac{a_2 - b_2}{a_2 b_2}\alpha_i$$

α_i and $\beta_i, i = 1, 2$ can be estimated. Thus, we get the following theorem:

Theorem 3.5.5 *Generalized Axial Design of Type-II ensures estimation of the parameters of the Draper–St. John's model* (3.5.18).

3.6 Concluding Remarks

For the estimation of the parameters in canonical homogeneous quadratic model (3.1.6), we have exploited the model expectations of suitably chosen experimental points, as indicated in Remark 3.2.1. It has been noted that for each of the designs, the information matrix has the desirable property that the elements are invariant with respect to permutation of the components.

Again we have seen that the generalized axial designs of different types play a good role in the estimation of the parameters in the quadratic and other types of models. The status of these designs may be examined in respect of optimality. However, it may be noted that Loewner order comparison of even a small subset of competing designs in the quadratic case may be quite difficult to handle.

References

Atwood, C. L. (1969). Optimal and efficient designs of experiments. *The Annals of Mathematical Statistics*, *40*, 1570–1602.

Cornell, J. A. (2002). *Experiments with mixtures* (3rd ed.). New York: Wiley.

Draper, N. R., & Pukelsheim, F. (1999). Kiefer ordering of simplex designs for first and second degree mixture models. *Journal of Statistical Planning and Inference*, *79*, 325–348.

Galil, Z., & Kiefer, J. (1977). Comparison of simplex designs for quadratic mixture models. *Technometrics*, *19*, 445–453.

Kiefer, J. (1961). Optimum designs in regression problems II. *The Annals of Mathematical Statistics*, *32*, 298–325.

Kiefer, J. (1975). Optimum design: Variation in structure and performance under change of criterion. *Biometrika*, *62*, 277–288.

Pal, M., Mandal, N. K., & Aggarwal, M. L. (2011). Optimum designs for parameter estimation in linear mixture model with synergistic effects. *Communications in Statistics-Theory and Methods*, *40*, 2059–2073.

Pal, M., & Mandal, N. K. (2012). Optimum designs for estimation of parameters in a quadratic mixture-amount model. *Communications in Statistics-Theory and Methods*, *41*, 665–673.

Quenouille, M. H. (1953). *The design and analysis of experiments*. London: Charles Griffin and Company.

Scheffé, H. (1958). Experiments with mixtures. Journal of the Royal Statistical Society. *Series B (Methodological)*, *20*, 344–360.

Scheffé, H. (1963). Simplex-centroid design for experiments with mixtures. Journal of the Royal Statistical Society. *Series B (Methodological)*, *25*, 235–263.

Sinha, B. K., Das, P., Mandal, N. K., & Pal, M. (2010). Parameter estimation in linear and quadratic mixture models: A review. *Pakistan Journal of Statistics*, *26*, 77–96.

Chapter 4
Optimal Mixture Designs for Estimation of Natural Parameters in Scheffé's Models

Abstract In this chapter, we review the optimality results for the estimation of parameters and subset of parameters of the Scheffé's mixture models while the factor space is the entire simplex, i.e., while we are in the framework of an unconstrained factor space. Though most of the results are related to quadratic model, optimality results for linear- and higher-degree polynomials are also discussed. Kiefer's equivalence theorem plays an important role in characterizing optimum designs. It has been observed that the support points of an optimum design under different optimality criteria belong to the subclass of union of barycenters.

Keywords Scheffé's mixture models · Estimation of parameters and subset of parameters · Specific optimum designs · Kiefer's equivalence theorem · Support points · Barycenters

4.1 Introduction

In a mixture experiment, the mean response depends on the proportions of the components in the mixture. If there be a totality of q (distinct) components in a mixture with the mixing proportions $(x_i, \ldots, x_q)$, then $x = (x_1, \ldots, x_q)' \in \mathcal{X}$, where

$$\mathcal{X} = \left\{ (x_1, \ldots, x_q) : x_i \geq 0, \quad i = 1, 2, \ldots, q; \sum_{i=1}^{q} x_i = 1 \right\}. \tag{4.1.1}$$

As mentioned in Chap. 2, Scheffé (1958) first introduced models in canonical forms of different degrees to represent the mean response function denoted by $\eta(x; \beta)$. The experimental region and the response functions in mixture experiments differ from the ordinary response surface problem in view of the constraint (4.1.1). Scheffé (1958, 1963) also introduced *simplex lattice designs* and *simplex centroid designs* appropriate in such situations. In Chap. 3, these have been introduced in details.

B. K. Sinha et al., *Optimal Mixture Experiments*, Lecture Notes in Statistics 1028,
DOI: 10.1007/978-81-322-1786-2_4, © Springer India 2014

In this chapter, we will study optimality aspect of mixture designs in a *continuous design* setting. For ready reference, this is again discussed below.

As mentioned in Chap. 3, a continuous design is represented by

$$\xi = \{\boldsymbol{x}_1, \boldsymbol{x}_2, \ldots, \boldsymbol{x}_n; p_1, p_2, \ldots, p_n\}, \tag{4.1.2}$$

which assigns mass $p_1, p_2, \ldots, p_n; p_i \geq 0, \sum p_i = 1$, to a collection of n support points $\boldsymbol{x}_1, \boldsymbol{x}_2, \ldots, \boldsymbol{x}_n$ of the design ξ in $\mathcal{X}$. In application, one could start with more than the desired number of support points from optimality point of view. In the final analysis, some of these points might be dropped. That is why, we have kept the possibility of $p_i = 0$ in (4.1.2).

In the following, we will be discussing optimality results with reference to different mixture models. It turns out that some standard mixture designs such as simplex/simplex centroid/axial designs play significant roles in such investigations. These designs have already been introduced and their salient features discussed in Chap. 3.

4.2 Optimum Designs for First and Second Degree Models

4.2.1 D- and I-Optimum Designs

Kiefer (1961) first studied the optimality of designs in the most general setting of (4.1.2) with reference to the problem of estimation of parameters of the Scheffé's model of degree one, two, and three (Vide Eqs. (3.1.2)–(3.1.5) of Chap. 3). The saturated design ξ which assigns measure $1/q$ to each of the q vertices of the simplex is D-optimum for Scheffé's canonical linear model in q components. Since $M(\xi) = I_q$, the design is ϕ_p-optimal as well for all $p > 0$ (Galil and Kiefer 1977). For the second-degree canonical model in q components, Kiefer (1961) also established D-optimality of $(q, 2)$ simplex lattice design, which puts equal mass at the support points of the design. He proved the optimality of the designs with the help of the equivalence theorem due to Kiefer and Wolfowitz (1960). This theorem is of utmost importance in the entire exercise on optimality by and large. We narrate below the essential features of this theorem.

Consider the general linear model,

$$y(\boldsymbol{x}) = \boldsymbol{f}'(\boldsymbol{x})\boldsymbol{\beta} + e(\boldsymbol{x}), \tag{4.2.1}$$

with usual assumptions on error component $e(\boldsymbol{x})$, viz. mean zero and uncorrelated homoscedastic variance σ^2.

The information matrix for $\boldsymbol{\beta}$, using a design ξ of the type (4.1.2) [in conjunction with (4.1.1)], is given by

$$M(\boldsymbol{\beta}, \xi) = \sum_{i=1}^{n} p_i \boldsymbol{f}(\boldsymbol{x}_i)\boldsymbol{f}'(\boldsymbol{x}_i).$$

Note that in the above expression, summation is over all the suffixes admitting of only positive mass. We assume estimability of the model parameters under the design ξ in $\mathcal{X}$.

Writing $d(\boldsymbol{x}, \xi) = \boldsymbol{f}'(\boldsymbol{x})M^{-1}(\boldsymbol{\beta}, \xi)\boldsymbol{f}(\boldsymbol{x})$, the celebrated equivalence theorem (Kiefer and Wolfowitz 1960) can be stated as follows

Theorem 4.2.1 *A design ξ^* is D-optimum for the estimation of $\boldsymbol{\beta}$ in* (4.2.1) *if and only if it satisfies $d(\boldsymbol{x}, \xi^*) \leq p$, for all $\boldsymbol{x}$ in $\mathcal{X}$ where p is the number of parameters in the model, with equality holding at all the support points of the design ξ^*.*

We readily refer to Scheffé models given in Chap. 3. The second-degree Scheffé model, in canonical form, [cf. Eq. (3.1.3) of Chap. 3] is given by

$$f(\boldsymbol{x}; \boldsymbol{\beta}) = \sum_{i=1}^{q} \beta_i x_i + \sum_{i=1}^{q} \sum_{j>i}^{q} \beta_{ij} x_i x_j \tag{4.2.2}$$

and, using the relation $\sum_{i=1}^{q} x_i = 1$, (4.2.2) can be further expressed as

$$f(\boldsymbol{x}; \boldsymbol{\beta}) = \sum_{i=1}^{q} \beta_{ii}^* x_i^2 + \sum_{i=1}^{q} \sum_{j>i}^{q} \beta_{ij}^* x_i x_j. \tag{4.2.3}$$

Clearly, β_{ij}^*'s are linear functions of the parameters of (4.2.2). Keeping in mind the support points of the $(q, 2)$ simplex design ξ^*, Kiefer (1961) expressed the second-degree model (4.2.3) in the following form:

$$f(\boldsymbol{x}; \boldsymbol{\beta}) = \sum_{i=1}^{q} \theta_{ii} x_i \left(x_i - \frac{1}{2}\right) + \sum_{i=1}^{q} \sum_{j>i}^{q} \theta_{ij} x_i x_j \tag{4.2.4}$$

$$= f^{*\prime}(\boldsymbol{x})\boldsymbol{\theta}, \text{ say,} \tag{4.2.5}$$

where θ_{ij} come out to be one-to-one functions of β_{ij}^*s and f^* is defined accordingly. He then established the D-optimality of $(q, 2)$ simplex lattice design for the estimation of the parameters of second-degree model.

After a little algebra, it can be readily shown that

$$\begin{aligned} d(\boldsymbol{x}, \xi^*) &= f^{*\prime}(\boldsymbol{x}) M^{-1}(\boldsymbol{\theta}, \xi^*) f^*(\boldsymbol{x}) \\ &= 4p\left[\sum_{i=1}^{q} x_i^2\left(x_i - \frac{1}{2}\right)^2 + 4\sum_{i=1}^{q}\sum_{j>i}^{q} x_i^2 x_j^2\right] \\ &= p\left[1 - \sum_{i \neq j} x_i x_j \left\{2\left(x_i - x_j\right)^2 + \left(1 - 4x_i x_j\right)\right\}\right] \\ &\leq p. \end{aligned} \tag{4.2.6}$$

Equality in (4.2.6) holds at the support points of ξ^*, the $(q, 2)$ simplex lattice design, which has the support points as indicated in Sect. 3.2 in Chap. 3. Hence, ξ^* is D-optimal. It is easy to check that designs different from $(q, 2)$ simplex lattice designs do not satisfy the equality condition above in (4.2.6) for all points in its support. Further, it is also to be noted that ξ^* being a $(q, 2)$ simplex lattice design is a saturated design.

As is well known, the D-optimality criteron is invariant with respect to linear transformations of the parameters of interest. That is why the parameters in the model (4.2.2)–(4.2.3) were recast in a different form in (4.2.4). There is more to it. The suggested alternative representation leads to a diagonal matrix for the information matrix when it is evaluated for the design ξ^*. This helps tremendously in the evaluation of $d(\boldsymbol{x}, \xi) = f'(\boldsymbol{x})M^{-1}(\boldsymbol{\beta}, \xi)f(\boldsymbol{x})$ for ξ^*. Kiefer succeeded in coming up with this crucial observation. The rest were routine task. The optimal design is, as we see, saturated and invariant. Also it has the special feature of producing all off-diagonal elements of the information matrix as zeros. This is a very powerful technique in the evaluation of optimum designs. As we will see in subsequent chapters, this technique has been exploited a number of times in determining optimum designs for the estimation of some nonlinear functions of the parameters.

Lambrakis (1968) considered a class of designs which assigns a weight (mass) p_1 to each $\boldsymbol{x} \leftrightarrow (1/2, 1/2(q-1), \ldots, 1/2(q-1))$ and a weight (mass) p_2 to each $\boldsymbol{x} \leftrightarrow (1/3, 1/3, 1/3\,(q-2), \ldots, 1/3(q-2))$. Optimum values of p_1/p_2 were obtained by using *I-optimality* criterion i.e., minimizing the expected variance of the predicted response over $\mathcal{X}$. Similarly, optimum values of p_1/p_2 were obtained when the points $\boldsymbol{x} \leftrightarrow (0, 1/(q-1), \ldots, 1/(q-1))$ and $\boldsymbol{x} \leftrightarrow (1/2, 1/2, 0, \ldots, 0)$ were used as support points (Lambrakis 1969). In the above, we have conveniently used the notation $\boldsymbol{x} \leftrightarrow$ (.....) to represent all distinct permutations of the elements of (....). Note that in Chap. 3 as well, this notation could be used to save space. Laake (1975) showed that a $\{q, 2\}$ simplex centroid design is I-optimal if $p_1/p_2 = [(q^2 - 7q + 18)/32]^{\frac{1}{2}}$.

4.2.2 ϕ_p-Optimum Designs

In this subsection, we propose to review some results on the optimality with respect to matrix means for mixture models.

Kiefer (1975) tried to characterize the ϕ_p-optimal design (also called criterion of matrix means) (see Kiefer 1974, 1975). He observed that the computation of a ϕ_p-optimal design is algebraically intractable; there is no hope for simple formulae. Hence, one must be satisfied with numerical results. However, the direct minimization is too large a problem to handle. Instead, he used various theoretical tools of optimal design theory to reduce the dimensionality of the problem. First using the convexity of the criterion function, he established that an optimum design is invariant under permutation of components. Using the equivalence theorem for ϕ_p-optimality criterion, he observed that the support points of an optimal design must be supported on a subset of $J = \cup J_j$, where J_j is a *barycentre* of depth j.

Galil and Kiefer (1977) extensively discussed the optimality of designs for the second-degree Scheffé model when the criterion function is one of the *matrix means*. They established that the optimal design must be supported by the collection of the centroids of the faces of the simplex and must be invariant under permutation of the coordinates of the support points. No other points beyond those described above will be involved. Galil and Kiefer (1977) used a Broyden minimization technique (Broyden 1967) to obtain numerical values of p_i at which the weighted $\{q, q\}$ centroid design is ϕ_p-optimum for $p = 0, 0, 1, 0, 5, 1, 2, 4, 5, 7, 10$ and $q = 2, 3, \ldots, 11$. The case of $p = \infty$ (E-optimality) was extensively studied. Computational routines for obtaining these designs are developed, and the geometry of structures is discussed. Except when $q = 3$, the A-optimum design is supported by the vertices and midpoints of edges of the simplex, as is the case for D-optimum design. They also showed that all of D-, A-, and E-optimum designs are reasonably robust in their efficiency under variation of optimality criteria, but E-optimum design is the most robust although it requires more support points. We will not pursue any serious presentation of the so-called optimality results under most general forms of matrix means criteria .

Kiefer (1978) discussed the asymptotics q-ingredients second-degree Scheffé models, as the number of ingredients tends to infinity. For the smallest eigen value criterion, formulas for the limiting weights, α_j of the elementary centroid designs η_j, also called barycentre of depth $j - 1$, are characterized. Kiefer then concentrated on the basic solutions in which four of the weights were positive. It emerged that α_1 and α_2 must always be positive, the remaining two weights then being α_j and α_h with $2 < j < h$. Formulas are given when $j = 2$, or $j = 3$.

Kiefer (1961) observed that the problem of determining D-optimum design for the estimation of a *subset of parameters* is not easy. For three-component mixture, he restricted his consideration to the coefficients of the mixed quadratic terms (vide 4.2.2). Then, he searched for the optimum design in the subclass of designs $\xi^{(\alpha)}$ which, for some α, assigns measure $\alpha/3$ to each vertex of $\mathcal{X}$ and measure $(1-\alpha)/3$ to the midpoint of each edge of $\mathcal{X}$. Denoting by $[a, b]$ *a* 3×3 matrix of the form

$$[a, b] = (a - b)I_3 + bJ_3$$

the information matrix underlying $\xi^{(\alpha)}$ can be expressed as

$$M(\xi^{(\alpha)}) = \frac{1}{3}\begin{pmatrix} [\frac{\alpha}{16}, 0] & [0, \frac{\alpha}{8}] \\ [0, \frac{\alpha}{8}] & [1 - \frac{\alpha}{2}, \frac{\alpha}{4}] \end{pmatrix}.$$

It can be readily verified that within the class of $\xi^{(\alpha)}$ designs, the unique D-optimal design corresponds to $\alpha = 0.6530$. Since x_1^2, x_2^2, and x_3^2 in the quadratic model, because of the constraints $\sum_i x_i = 1$, are related to x_1x_2, x_1x_3 and x_2x_3 by a non-singular transformation, a design which is D-optimum for the estimation of parameters associated to the quadratic terms is also D-optimum for the estimation of parameters related to the cross product terms and vice versa.

That the above characterization indeed leads to D-optimal in the whole class follows from a non-trivial application of the equivalence theorem, as appropriately modified and applied in the context of estimation of a subset of parameters. The basic results in this direction are due to Karlin and Studden (1966), followed by Atwood (1969). We will defer description and discussions on these results to subsequent chapters.

4.2.3 Kiefer-Optimal Designs

Draper and Pukelsheim (1999) discussed the problem of improvement of a given design in terms of (i) increasing *symmetry (nearness* of diagonal elements and also *nearness* of off-diagonal elements, each within the context of subgroups of elements) as well as (ii) obtaining a *larger* information matrix under *Loewner Ordering.* The two criteria together constitute the *Kiefer Design Ordering* (KDO). Symmetry has always been a prime attribute of good experimental designs and comprise the first step of the KDO. The second step is an improvement relative to the Loewner ordering within the class of exchangeable information matrices.

Draper and Pukelsheim (1999) mainly worked with *K-model* based on Kronecker algebra of vectors and matrices in contrast to *S-model* of Scheffé, introduced earlier in this chapter. The expected response (i.e., the mean model) under second-degree K-model takes the form $\eta(\boldsymbol{x}) = \sum_{i=1}^{q}\sum_{j=1}^{q} \theta_{ij}x_ix_j = (\boldsymbol{x} \otimes \boldsymbol{x})'\boldsymbol{\theta}$; see Draper and Pukelsheim (1998, 1999). It may be mentioned that their results on the Kiefer ordering of experimental designs for second-degree mixture models do not depend on the actual parameterization of the response function. Further, it may also be noted that the identifiability of the parameters are lost momentarily. Before citing results on KDO of information matrices, let us define *elementary centroid designs* η_j, for $j = 1, \ldots, q$. For a specified $j < q$, the design η_j rests on $C(q, j)$ points of the form $\boldsymbol{x} \leftrightarrow (1/j, 1/j, \ldots, 1/j, 0, 0\ \ldots, 0)$ and also it places equal weight $1/C(q, j)$ on each of these points. A simplex centroid design is a mixture (i.e., weighted union) of *all* elementary centroid designs η_j, for $j = 1, 2, \ldots, q$. (vide Chap. 3 for details.)

Once for all, it may be mentioned that below we are referring to K-models only.

For the first-degree model (in canonical form i.e., homogeneous model of degree 1), Draper and Pukelsheim (1999) established the following:

Theorem 4.2.1 *Let* $\bar{\tau}$ *be an invariant design in* (4.1.1). *Then*

$$M(\eta_1) \geq M(\bar{\tau})$$

with equality holding iff $\bar{\tau} = \eta_1$.

It is easy to see that $M(\eta_1) = (1/q)I_q$. Also we see that there is no essential difference between K-model and S-model when we are dealing with a first-degree model.

For the second-degree K-model, they restricted to two- and three-component mixtures and established the following theorem:

Theorem 4.2.2 *In the second-degree mixture model, the set of weighted centroid designs constitutes a minimal complete class of designs for the Kiefer Ordering.*

Afterward, Draper et al. (2000) extended this result to four and more components:

Theorem 4.2.3 *In the second-degree mixture model for* $q \geq 4$ *ingredients, the set of weighted centroid designs* $\mathcal{C} = \{\alpha_1\eta_1 + \alpha_2\eta_2 + \cdots + \alpha_q\eta_q;\ \alpha_i \geq 0, \sum \alpha_i = 1\}$ *is convex and constitutes a complete class of designs for the Kiefer ordering. For* $q = 4$, *the class is minimal complete.*

Further, they selected a subclass of the set $\mathcal{C}$ in Theorem 4.2.3 above that is essentially complete and established the following.

Theorem 4.2.4 *In the second-degree mixture model for five or more ingredients, the set* $\mathcal{C} = \bigcup_{j=2}^{q-2} conv\,\{\eta_1, \eta_j, \eta_{j+1}, \eta_q\}$ *constitutes an essentially complete class of exchangeable designs for the Kiefer ordering.*

In the above, *conv* refers to a convex combination of the constituent components. For further details, the readers are referred to the original papers cited above.

Draper and Pukelsheim (1998) and Prescott et al. (2002) put forward several advantages of the Kronecker model, e.g., the homogeneity of regression terms and reduced ill-conditioning of information matrices. Interestingly, both models share the same invariance properties. Because of the completeness Theorems 4.2.3–4.2.4 in Draper et al. (2000), the design problem reduces to the class of weighted centroid designs. Based on these observations, Klein (2004) investigated optimal designs in the second-degree Kronecker model for mixture experiments and presented the following: (i) characterization of feasible weighted centroid designs for a maximum parameter system, (ii) derivations of D-, A-, and E-optimal weighted centroid designs, and (iii) numerical computation of ϕ_p-optimal weighted centroid designs. Results on quadratic subspace of invariant symmetric matrices containing the information matrices involved in the design problem served as a main tool in his analysis.

4.3 Polynomial Models of Degree Three and More

The cases, where the degree of the polynomial representing the response function is more than 2, are computationally very difficult. For the *third-degree* model with $q = 2$ factors, the response function η_x, because of the constraint $x_1 + x_2 = 1$, reduces to a cubic polynomial in x_1. This is a trivial observation. Actually, one starts with model in canonical form

$$\eta_x = \beta_1 x_1 + \beta_2 x_2 + \beta_{12} x_1 x_2 + \alpha_{12} x_1 x_2 (x_1 - x_2)$$

which reduces to

$$\eta_x = \gamma_0 + \gamma_1 x_1 + \gamma_2 x_1^2 + \gamma_3 x_1^3,$$

because of $x_1 + x_2 = 1$, where $\boldsymbol{\gamma} = (\gamma_0, \gamma_1, \gamma_2, \gamma_3)'$ is related to $\boldsymbol{\beta} = (\beta_1, \beta_2, \beta_{12}, \alpha_{12})'$ through $\boldsymbol{\gamma} = T\boldsymbol{\beta}$ with

$$T = \begin{bmatrix} 0 & 1 & 0 & 0 \\ 1 & -1 & 1 & -1 \\ 0 & 0 & -1 & 3 \\ 0 & 0 & 0 & -2 \end{bmatrix},$$

and T is non-singular.

By classical argument, the D-optimum design will have four support points—two at the two extremes and two inside the domain. Again, since the problem is invariant with respect to the two components (both in terms of the representation and the D-optimality criterion), the points inside the domain must be of the form $(b, 1 - b)$ and $(1 - b, b)$ for some $b, 0 < b < 1$. Then, b is determined so as to optimize the D-optimality criterion, and it comes out as

$$b = b_0 = \frac{1}{2}\left(1 - \frac{1}{\sqrt{5}}\right).$$

Thus, the optimum design has the support points $(0, 1)$, $(1, 0)$, $(b_0, 1 - b_0)$ and $(1 - b_0, b_0)$, with equal masses. Note that *equal mass* consideration applies since the design is saturated. Because of the invariance property of the D-optimality criterion under non-singular transformation of the parameter vector, the above design is also D-optimum for the estimation of the original parameters β's.

For the case of three factors, Kiefer (1961) studied different cubic polynomials introduced by Scheffé (1958). In the model involving nine parameters and having nine functions of the form $x_i, x_ix_j, x_ix_j(x_i - x_j)$, but without the function $x_1x_2x_3$, he considered the design ξ_b, which puts positive mass at

(i) each of the three extreme points $(0, 0, 1), (0, 1, 0), (1, 0, 0)$, and
(ii) each of six points on the edges $(b, 1 - b, 0), (1 - b, b, 0), (b, 0, 1 - b), (1 - b, 0, b), (0, b, 1 - b), (0, 1 - b, b)$.

The mass distribution is confined to the above nine points only so that the design is saturated. Hence, for *D*-optimality, the mass is taken to be 1/9 for each support point. Then, it turns out that

$$\text{det.}M(\xi_b) \propto v^{12}(1 - 4v)^3,$$

where $v = b(1 - b)$.

The optimal value of v, which maximizes $\text{det.}M(\xi_b)$, comes out to be $v = \frac{1}{5}$, and hence, optimum b is

$$b = b_0 = \frac{1}{2}\left(1 - \frac{1}{\sqrt{5}}\right).$$

This solution to b_0 is strikingly the same as the earlier solution for the case of two components. Other contributors in this area include:

Farrel et al. (1967) considered the cubic model in three-component case with 10 functions, viz, $x_i, x_ix_j, x_ix_j(x_i - x_j)$ and $x_1x_2x_3$. Then, they showed that a design that assigns equal mass to the ten points consisting of three vertices $(1, 0, 0)$, $(0, 1, 0)$, and $(0, 0, 1)$, the simplex centroid point $(1/3, 1/3, 1/3)$ and the six permutations of $(b, 1 - b, 0)$ with $b = \frac{1}{2}(1 - \frac{1}{\sqrt{5}})$ is *D*-optimal. They verified the optimality of the design via equivalence theorem, using the orthogonal representation of the regression functions relative to the support points of the optimal design.

Remark 4.3.1 Lim (1990) extended the results of Farrell et al. (1967) on *D*-optimality for the third-degree model with 4 or more factors. Case $q = 4$ is solved conclusively, and the cases $q = 5, 6, \ldots, 10$ are solved numerically.

Kasatkin (1974) considered a design which assigns the same weight to each $x \leftrightarrow (1, 0, \ldots, 0)$, $x \leftrightarrow (\alpha, 1 - \alpha, 0, \ldots, 0)$ and $x \leftrightarrow (1/3, 1/3, 1/3, 0, 0)$ for estimation of coefficients in the full cubic model:

$$\eta_x = \sum_{i=1}^{q} \beta_i x_i + \sum_{i=1}^{q} \sum_{\substack{j=i \\ i<j}}^{q} \beta_{ij} x_i x_j + \sum_{i=1}^{q} \sum_{\substack{j=i \\ i<j}}^{q} \gamma_{ij} x_i x_j (x_i - x_j)$$
$$+ \sum_{i=1}^{q} \sum_{\substack{j=i \\ i<j}}^{q} \sum_{\substack{k=i \\ j<k}}^{q} \beta_{ijk} x_i x_j x_k$$

and obtained the optimum value α so that the determinant of the information matrix is a maximum. He obtained similar results for the fourth- and fifth-degree polynomial.

In general, for the general Scheffé (q, n) polynomial model

$$\eta_x = \sum_{i=1}^{q} \beta_i x_i + \sum_{1 \le i < i_2 \le q} \beta_{i_1 i_2} x_{i_1} x_{i_2} + \sum_{1 \le i_1 < i_2 < \cdots < i_n \le q} \beta_{i_1 i_2 \ldots i_n} x_{i_1} x_{i_2} \ldots x_{i_n} \quad (4.3.1)$$

it is difficult to derive for the estimation of the parameters. However, some theoretical and numerical results are available in the literature.

Atwood (1969) showed that for $4 \le n \le q$, the (q, n) simplex centroid design is D-optimal for the estimation of parameters of the model (4.3.1) when $n = q$ but fails to be so for $n < q$. Guan and Chao (1987) and Laake (1975) obtained A-optimal allocations for the weighted (q, n) simplex centroid designs for the model (4.3.1) with $n \le q$ while Guan (1988) obtained I-optimal allocation, Liu and Neudecker (1995) studied V-optimality of weighted (q, q) simplex centroid designs for the model parameters of (4.3.1) with $n = q$. Guan and Liu (1989, 1993) showed that D- and A-optimal designs for model (4.3.1) with $n = q - 1$ are weighted (q, q) centroid designs.

As for the case of double mixtures where two mixtures x and z are contained in a model, Lambrakis (1968) obtained the I-optimal allocation of points in a $\{p, q; r, s\}$ double lattice, which is produced from the $\{p, r\}$ and $\{q, s\}$ lattices, for the estimation of coefficients of the mean model $\eta_{p,q;r,s}$, which is of rth degree in x and sth degree in z where $r = 2$, $s = 3$. He claimed that this result was verified for $1 \le r \le 4$, $1 \le s \le 4$ and conjectured that it is also true for $r, s > 4$.

4.4 Mixture-Amount Model

In some situations, for instance in experiments on fertilizers, the response depends not only on the proportions of several ingredients, but also on the *total amount* of the mixture used. To accommodate the influence of the amount of mixture on the response, several models have been proposed. Piepel and Cornell (1985) introduced new mixture-amount models that explain the effect of amount on the blending properties of components by considering the regression coefficients of the usual mixture models, proposed by Scheffé (1958, 1963), as functions of amount. One type of such models which is quadratic in the total amount A for a quadratic mixture model can be expressed as

$$\eta_x = \sum_{k=0}^{2} \left(\sum_{i=1}^{q} \gamma_i^{(k)} x_i + \sum_{i=1}^{q-1} \sum_{j>i}^{q} \gamma_{ij}^{(k)} x_i x_j \right) A^k. \tag{4.4.1}$$

Remark 4.4.1 The above model can also be looked as a quadratic Scheffé model in canonical form where the coefficients of different terms are functions of the mixture amount. When such functions are quadratic in A, the Scheffé model reduces to (4.4.1) (see also Cornell 2002).

Such cross product type models are usually called *mixture-amount models* and can be generalized by using lower- or higher-order terms and different powers of A. This model is not generally appropriate when $A = 0$, because it predicts a zero response. Modifications that could be made to the model to take account of the zero-amount situation were discussed in Piepel (1988). An alternative model, which allows estimation of a zero response, can be fitted to the actual *amounts* of the ingredients rather than the *proportions*. Suppose that the maximum total amount, $A_{\max}$, is coded to be a and that the actual amounts of the q mixture ingredients are denoted by $a_1, a_2, \ldots, a_k$ so that $a_1 + a_2 + \cdots + a_q = A$, with $0 \leq A \leq A_{\max}$. The proportions x_i are related to the amount a_i through $a_i = x_i A$. A polynomial model in the a_i, called a *component-amount model*, may be used to represent the response. For a second-degree fit, this alternative model is the full quadratic.

$$\eta_x = \alpha_0 + \sum_{i=1}^{q} \alpha_i a_i + \sum_{i=1}^{q-1} \alpha_{ii} a_i^2 + \sum_{i=1}^{q-1} \sum_{j>i}^{q} \alpha_{ij} a_i a_j.$$

An excellent discussion of mixture-amount and component-amount models is given by Cornell (2002). Prescott and Draper (2004) considered component-amount designs formed from projections of simplex lattice and simplex centroid designs into lower-dimensional spaces. Afterward, they extended the above ideas by finding the *D*-optimal designs contained in this class of projection designs.

Hilgers and Bauer (1995) obtained optimum designs for estimation of parameters in tic-type polynomial mixture-amount modes. Heiligers and Hilgers (2003) established a close relation between *admissible* mixture and *admissible* mixture-amount designs in additive and homogeneous models. This particularly allows to obtain *D*-, *A*- and *V*-optimal mixture-amount design from optimal mixture designs, and vice versa. The authors also presented some examples for Becker's and Scheffé's mixture models.

Zhang et al. (2005) attempted to find *D*- and *A*-optimum designs for parameter estimation in quadratic Darroch-Waller model (1985), extended to include amount. They showed that the origin and vertices of the simplex are the support points of the optimal designs, and when the number of mixture components increases, other support points shift gradually from barycentres of depth 1 to barycentres of higher depths. Pal and Mandal (2012) introduced the following mixture-amount model which is quadratic, both in the amount and the proportions and obtained A- and D- optimal design for the estimation of the model parameters:

$$\eta_{\mathbf{x}} = \kappa_{00}^* + \kappa_{01}^* A + \kappa_{02}^* A^2 + A \sum_{i=1}^{q} \alpha_{0i}^* x_i + \sum_{i=1}^{q} \alpha_{ii}^* x_i^2 + \sum_{j>i}^{q} \alpha_{ij}^* x_i x_j \tag{4.4.2}$$

where $A \in [A_L, A_U], A_L > 0$, denotes the amount and $x_1, x_2, \ldots, x_q$ denote the proportions of the q components in the mixture. Clearly, $x_i \geq 0$, for $i = 1, 2, \ldots, q$ and $\sum_{i=1}^{q} x_i = 1$. The assumption $A_L > 0$ ensures that some amount of the mixture should necessarily be used in the experiment.

Model (4.4.2) finds usefulness in situations, where synergism is present between the amount and the mixing proportions. This is essentially true when any change in the amount of the mixture affects the blending properties of the mixture components, besides causing change in the response. The difference between this model and that suggested by Piepel and Cornell (1985) is that while Piepel and Cornell (1985) explained the effect of amount on the mixing proportions by taking the regression coefficients to be functions of amount, in model (4.4.2) the interactions between amount and mixing proportions explain this effect. Further, in comparison to the model suggested by Piepel and Cornell (1985), model (4.4.2) has fewer parameters.

By *suitable transformation on the amount*, and using the restriction $\sum_{i=1}^{q} x_i = 1$, (4.4.2) can be written as

$$\eta_x = \kappa_{01} A + \kappa_{02} A^2 + A \sum_{i=1}^{q} \alpha_{0i} x_i + \sum_{i=1}^{q} \alpha_{ii} x_i^2 + \sum_{i<j=1}^{q} \alpha_{ij} x_i x_j \tag{4.4.3}$$

where κ_{0i}s and α_{ij}s are linear functions of κ_{0i}^*s and α_{ij}^*s, and the factor space is given by

$$\mathcal{X}_A = \{(A, x_1, x_2, \ldots, x_q) : A \in [-1, 1], x_i \geq 0, 1 \leq i \leq q, \sum_{i=1}^{q} x_i = 1\}. \tag{4.4.4}$$

Because of the constraint $\sum_{i=1}^{q} x_i = 1$, (4.4.3) can again be represented in the form

$$n_x = \beta_{00} A^2 + A \sum_{i} \beta_{0i} x_i + \sum_{i} \beta_{ii} x_i^2 + \sum_{i<1} \beta_{ij} x_i x_j \tag{4.4.5}$$

where

$$\beta_{00} = \alpha_{02}, \quad \beta_{0i} = \alpha_{01} + \alpha_{0i}, \ 1 \leq i \leq q,$$
$$\beta_{ii} = \alpha_{ii}, \ 0 \leq i \leq q, \quad \beta_{ij} = \alpha_{ij}, \ 1 \leq i < j \leq q.$$

Let, $\boldsymbol{x}^* = (A, x_1, x_2, \ldots, x_q)'$. Then, $\boldsymbol{x}^*$ satisfies the constraint

$$\boldsymbol{c}'\boldsymbol{x}^* = 1, \tag{4.4.6}$$

where $\boldsymbol{c} = (0, 1, 1, \ldots, 1)'$.

The problem considered by Pal and Mandal (2012) is to find a (continuous) design in the factor space (4.4.4) so that β_{ij}s in model (4.4.4) can be estimated with maximum accuracy.

Using a design ξ, one can estimate the parameters of the model (4.4.5). The information (moment) matrix of the design is given by

$$M(\xi) = \sum p_i \boldsymbol{f}(\boldsymbol{x}_i^*)\boldsymbol{f}(\boldsymbol{x}_i^*)', \tag{4.4.7}$$

where
$\boldsymbol{f}(\boldsymbol{x}_i^*) = (A_i^2, A_i x_{i1}, A_i x_{i2}, \ldots, A_i x_{iq}, x_{i1}^2, x_{i2}^2, \ldots, x_{iq}^2, x_{i1}x_{i2}, \ldots, x_{i,q-1}x_{iq})'$.

Design optimality aims at minimizing some function of $M^{-1}(\xi)$, or maximizing some function of $M(\xi)$. For comparing different designs, Pal and Mandal (2012) considered the A-optimality and D-optimality criteria, where the criterion functions for minimization are given by

$$\phi_A(\xi) = \text{Trace}(M^{-1}(\xi))$$
$$\phi_D(\xi) = \text{Det.}(M^{-1}(\xi)). \tag{4.4.8}$$

To obtain the optimal designs, note that both $\phi_A(\xi)$ and $\phi_D(\xi)$ are invariant with respect to the proportions. Hence, the optimum design will also be invariant with respect to x_is. As such, in respect of x_is, one may confine to $(q, 2)$-simplex centroid designs with equal weights at the vertices and the midpoint of the edges, respectively. Further, for x_is given, since the model (4.4.5) is quadratic in A, the optimum design is likely to admit three distinct values of A, two at the two extremes and one in between, with positive weights. Hence, the authors initially confine their search for A-optimal design within the subclass $\mathcal{D}_q$ of designs having the support points and weights as given in Table 4.1.

Let, $0 \leq p_i, p_i', p_i'' \leq 1, i = 1, 2, C(q, 1)p_1 + C(q, 2)p_2 = 1, C(q, 1)p_1' + C(q, 2)p_2' = 1, C(q, 1)p_1'' + C(q, 2)p_2'' = 1, a_0 \in (-1, 1), w_j \geq 0, j = -1, 0, 1$ and $w_{-1} + w_0 + w_1 = 1$, w_{-1}, w_0 and w_1 denote the weights attached to $A = -1, a_0, 1$ respectively, while the sixth column in Table 4.1 gives the weights for different $(x_1, x_2, \ldots, x_q)$ combinations when A is given.

Table 4.1 The subclass of designs D_q

x_1	x_2	...	x_{q-1}	x_q	Weight	A	Weight
1	0	...	0	0	p_1	-1	w_{-1}
0	1	...	0	0	p_1		
0	0	...	1	0	p_1		
0	0	...	0	1	p_1		
1/2	1/2	...	0	0	p_2		
1/2	0	...	0	0	p_2		
0	0	...	0	1/2	p_2		
0	0	...	1/2	1/2	p_2		
1	0	...	0	0	p'_1	a_0	w_0
0	1	...	0	0	p'_1		
0	0	...	1	0	p'_1		
0	0	...	0	1	p'_1		
1/2	1/2	...	0	0	p'_2		
1/2	0	...	0	0	p'_2		
0	0	...	0	1/2	p'_2		
0	0	...	1/2	1/2	p'_2		
1	0	...	0	0	p''_1	1	w_1
0	1	...	0	0	p''_1		
0	0	...	1	0	p''_1		
0	0	...	0	1	p''_1		
1/2	1/2	...	0	0	p''_2		
1/2	0	...	0	0	p''_2		
0	0	...	0	1/2	p''_2		
0	0	...	1/2	1/2	p''_2		

For any design $\xi \in D_q$, the moment matrix is given by

$$M(\xi) = \begin{bmatrix} M_{11} & M_{12} & M_{13} \\ M'_{12} & M_{22} & M_{23} \\ M'_{13} & M'_{23} & M_{33} \end{bmatrix},$$

where

$$M_{11} = \begin{bmatrix} w_1 + w_{-1} + a_0^4 w_0 & a_{11} 1'_q \\ & b_{11} I_q + c_{11} 1_q 1'_q \end{bmatrix},$$

$$M_{12} = \begin{bmatrix} a_{12} 1'_q \\ b_{12} I_q + c_{12} 1_q 1'_q \end{bmatrix}, \quad M_{13} = \begin{bmatrix} a_{13} 1' c(q,2) \\ c_{13} M_0 \end{bmatrix},$$

$$M_{22} = b_{22} I_q + c_{22} 1_q 1'_q, \quad M_{23} = c_{23} M_0, \quad M_{33} = b_{33} I c(q,2)$$

$$
\begin{aligned}
a_{11} &= \left\{p_1 + \frac{q-1}{2} p_2\right\} w_1 - \left\{p_1 + \frac{q-1}{2} p_2''\right\} w_{-1} + a_0^3 \left\{p_1' + \frac{q-1}{2} p_2'\right\} w_0 \\
b_{11} &= \left\{p_1 + \frac{q-2}{2} p_2\right\} w_1 + \left\{p_1 + \frac{q-2}{2} p_2''\right\} w_{-1} + a_0^2 \left\{p_1' + \frac{q-2}{2} p_2'\right\} w_0 \\
c_{11} &= \frac{p_2}{4} w_1 + \frac{p_2''}{4} w_{-1} + a_0^2 \frac{p_2'}{4} w_0 \\
a_{12} &= \left\{p_1 + \frac{q-1}{4} p_2\right\} w_1 + \left\{p_1'' + \frac{q-1}{4} p_2\right\} w_{-1} + a_0^2 \left\{p_1' + \frac{q-1}{4} p_2'\right\} w_0 \\
b_{12} &= \left\{p_1 + \frac{q-2}{8} p_2\right\} w_1 - \left\{p_1'' + \frac{q-2}{2} p_2''\right\} w_{-1} + a_0 \left\{p_1' + \frac{q-2}{8} p_2'\right\} w_0 \\
c_{12} &= \frac{p_2}{8} w_1 - \frac{p_2''}{8} w_{-1} + a_0 \frac{p_2'}{8} w_0 \\
a_{13} &= 2\left\{\frac{p_2}{8} w_1 - \frac{p_2''}{8} w_{-1} + a_0^2 \frac{p_2'}{8} w_0\right\} \\
c_{13} &= \frac{p_2}{8} w_1 - \frac{p_2''}{8} w_{-1} + a_0 \frac{p_2'}{8} w_0 \\
b_{22} &= \left\{p_1 + \frac{q-2}{16} p_2\right\} w_1 - \left\{p_1'' + \frac{q-2}{16} p_2''\right\} w_{-1} + \left\{p_1' + \frac{q-2}{16} p_2'\right\} w_0 \\
c_{22} &= \frac{p_2}{16} w_1 + \frac{p_2''}{16} w_{-1} + \frac{p_2'}{16} w_0 = c_{23} = b_{33},
\end{aligned}
$$

M_0 is a $q \times C(q, 2)$ matrix in which the first $q - 1$ elements in the first row are 1 and the remaining are 0, and the following $q - 1$ rows are permutations of the first row, $\mathbf{1}_s$ is a $s \times 1$ vector with all components unity and $\mathbf{I}_s$ denotes an identity matrix of order s.

Since for quadratic regression in $[-1, 1]$, the optium support points of D-, A-, and E-optimality criteria are at $-1, 0$, and 1, to start with, the authors take $a_0 = 0$, and $w_{-1} = w_1, p_i = p_i'', i = 1, 2$. Let, $\mathcal{D}_q^0 \subset \mathcal{D}_q$ define the corresponding subclass of designs.

Then, for any design $\xi \in \mathcal{D}_q^0$, $M(\xi)$ has the representation

$$
\begin{array}{cccc}
A^2 & Ax_1 \; Ax_2 \; \ldots \; Ax_q & x_1^2 \; x_2^2 \; \ldots \; x_q^2 & x_1x_2 \; x_1x_3 \; \ldots \; x_{q-1}x_q
\end{array}
$$

$$
\begin{bmatrix}
2w_1 & \mathbf{0}_q' & 2w_1(p_1 + \frac{q-1}{4} p_2)\mathbf{1}_q', & 2w_1 \frac{p_2}{4} \mathbf{1}_{C(q,2)}' \\
 & 2w_1\{(p_1 + \frac{p-2}{4} p_2)I_q + \frac{p_2}{4} \mathbf{1}_q \mathbf{1}_q'\} & 0 & 0 \\
 & & r_1 I_q + r_2 \mathbf{1}_q \mathbf{1}_q' & r_2 M_0 \\
 & & & r_2 I_{C(q,2)}
\end{bmatrix},
$$

where

$$r_1 = 2w_1\left(p_1 + \frac{q-2}{16}p_2\right) + (1-2w_1)\left(p_1' + \frac{q-2}{16}p_2'\right),$$
$$r_2 = 2w_1\frac{1}{16}p_2 + (1-2w_1)\frac{1}{16}p_2'.$$

And,

$$M^{-1}(\xi) = \begin{bmatrix} M_{11}^* & M_{12}^* \\ M_{21}^* & M_{22}^* \end{bmatrix}$$

where

$$M_{11}^* = \frac{1}{2w_1}\begin{bmatrix} \frac{1}{1-2w_1k_1} & \mathbf{0}' \\ \mathbf{0} & k_2 I_q + k_3 1_q 1_q' \end{bmatrix}, M_{12}^* = -\frac{1}{(1-2w_1k_1)}\begin{bmatrix} t1_q' & s1_{C(q,2)}' \\ \mathbf{0} & \mathbf{0} \end{bmatrix}$$
$$M_{22}^* = \begin{bmatrix} \frac{1}{r}I_q & -\frac{1}{r}M_0 \\ -\frac{1}{r}M_0 & \frac{1}{r_2}I_{C(q,2)} + \frac{1}{r}M_0'M_0 \end{bmatrix} + \frac{2w_1}{1-2w_1k_1}\begin{bmatrix} t^2 1_q 1_q' & ts1_q 1_{C(q,r)}' \\ st1_{C(q,2)}1_q' & s^2 1_{C(q,2)}1_{C(q,2)}' \end{bmatrix}$$
$$k_1 = \frac{q}{r}p_1^2 + \frac{C(q,2)}{r_2}\frac{p_2^2}{16}, k_2 = \frac{1}{p_1 + \frac{q-2}{4}p_2}$$
$$k_3 = -\frac{p_2}{4(p_1 + \frac{q-2}{4}p_2)(p_1 + \frac{q-1}{2}p_2)}$$
$$t = \frac{p_1}{r}, s = -\frac{2}{r}p_1 + \frac{p_2}{4r_2}, r = 2w_1p_1 + (1-2w_1)p_1'.$$

Then,

$$\phi_A(\xi) = \frac{1}{2w_1(1-2w_1k_1)} + \frac{q}{2w_1}(k_2+k_3) + \frac{q}{r} + C(q,2)\left(\frac{1}{r_2} + \frac{2}{r}\right) + \frac{2w_1}{1-2w_1k_1}\{qt^2 + C(q,2)s^2\}. \quad (4.4.9)$$

The optimal values of p_1, p_1', and w_1 are obtained by minimizing $\phi_A(\xi)$. The authors checked the optimality or otherwise of the design within the entire class using Kiefer's equivalence theorem (1974), which, for A-optimality criterion, reduces to the following (see Pal and Mandal 2007):

Theorem 4.4.1 *A necessary and sufficient condition for a mixture-amount design ξ^* to be A-optimal within the whole class of competing designs is that*

$$\boldsymbol{f}'(\boldsymbol{x}^*)M^{-2}(\xi^*)\boldsymbol{f}(\boldsymbol{x}^*) \le trM^{-1}(\xi^*) \quad (4.4.10)$$

holds for all $\boldsymbol{x}^* \in \mathcal{X}_A$. Equality in (4.4.10) *holds at the support points of ξ^*.*

Table 4.2 *A*-optimal designs in q-component mixture-amount model for $q = 2, 3, 4, 5$

q	p_1	p'_1	w_1	w_0
2	0.2939	0.1571	0.2599	0.4802
3	0.1502	0.0505	0.2750	0.4501
4	0.0963	0.0205	0.2873	0.4254
5	0.0690	0.0081	0.2977	0.4046

Table 4.3 *D*-optimal designs in q-component mixture-amount model for $q = 2, 3, 4, 5$

q	p_1	p'_1	w_1	w_0
2	0.3925	0.3121	0.3717	0.2566
3	0.2146	0.1163	0.3936	0.2128
4	0.1362	0.0372	0.4079	0.1842
5	0.0941	0.0011	0.4184	0.1632

As the algebraic derivations are rather involved, the condition (4.4.10) is checked by numerical computation. Table 4.2 gives *A*-optimal designs for $2 \le q \le 5$ which have been numerically examined by the authors to satisfy condition (4.4.10).

The *D*-optimality criterion comes out by maximizing $\phi_D(\xi)$, derived directly from the moment matrix, rather than its inverse, where

$$\phi_D(\xi) = (2w_1)^{q+1}(1 - 2w_1k_1)\left\{p_1 + 2(q-1)\frac{p_2}{4}\right\}\left\{p_1 + (q-2)\frac{p_2}{4}\right\}^{q-1} r_2^{C(q,2)} \times \{r_1 - (q-2)r_2\}^q. \tag{4.4.11}$$

The values of p_1, p'_1, and w_1 in the *D*-optimal design maximize (4.4.11). Optimality of the designs in the entire class of competing designs is checked using the Kiefer's equivalence theorem, which, for *D*-optimality criterion in the present setup, is as follows:

Theorem 4.4.2 *A necessary and sufficient condition for a mixture-amount design* ξ^* *to be D-optimal within the whole class of competing designs is that*

$$\boldsymbol{f}'(\boldsymbol{x}^*)M^{-1}(\xi^*)\boldsymbol{f}(\boldsymbol{x}^*) \le \frac{(q+1)(q+2)}{2} \tag{4.4.12}$$

holds for all $\boldsymbol{x}^* \in \mathcal{X}_A$.

Equality in (4.4.12) *holds at the support points of* ξ^*.

Owing to the complexity in algebraic derivation, validity of (4.4.12) is checked by numerical computation for $2 \le q \le 5$ (Table 4.3).

For the full set of natural mixture model parameters as also for a subset of parameters, we have presented/discussed results on optimal mixture designs. This study is based on the assumption that the factor space is the entire simplex without any constraints.

References

Atwood, C. L. (1969). Optimal and efficient designs of experiments. *The Annals of Mathematical Statistics*, *40*, 1570–1602.

Broyden, C. G. (1967). Quasi Newton methods and the application to function minimization. *Mathematics of Computation*, *21*, 368–381.

Cornell, J. A. (2002). *Experiments with mixtures* (3rd ed.). New York: Wiley.

Darroch, J. N., & Waller, J. (1985). Additivity and interaction in three-component experiments with mixtures. *Biometrika*, *72*, 153–163.

Draper, N. R., & Pukelsheim, F. (1998). Mixture models based on homogeneous polynomials. *Journal of Statistical Planning and Inference*, *71*, 303–311.

Draper, N. R., & Pukelsheim, F. (1999). Kiefer ordering of simplex designs for first and second degree mixture models. *Journal of Statistical Planning and Inference*, *79*, 325–348.

Draper, N. R., Heiligers, B., & Pukelsheim, F. (2000). Kiefer ordering of simplex designs for second-degree mixture models with four or more ingredients. *Annals of Statistics*, *28*, 578–590.

Farrel, R. H., Kiefer, J., & Walbran, A. (1967). Optimum multivariate designs. *Proceedings of the fifth Berkeley symposium on mathematical statistics and probability* (pp. 113–138). Berkeley: University of California.

Galil, Z., & Kiefer, J. (1977). Comparison of simplex designs for quadratic mixture models. *Technometrics*, *19*, 445–453.

Guan, Y., & Chao, X. (1987). On the A-optimal allocation of observations for the generalized simplex centroid designs. *Journal of Engineering Mathematics*, *4*, 33–39. (in Chinese).

Guan, Y. (1988). The I_λ-optimal allocation of observation in a generalized simplex-centroid design. *Chinese Journal of Applled Probability Statistics*, *4*, 157–162. (in Chinese).

Guan, Y., & Liu, C. (1989). D-optimal design for multi-linear polynomial model of q-1 degree (in Chinese). *Mathematics Statistics Applications Probability*, *4*, 353–361.

Guan, Y., & Liu, C. (1993). A-optimal design for multi-linear polynomial model of q-1 degree. *Journal of Northeastern University*, *14*, 198–202. (in Chinese).

Heiligers, B., & Hilgers, R. D. (2003). A note on optimum mixture and mixture amount designs. *Statistica Sinica*, *13*, 709–725.

Hilgers, R. D., & Bauer, P. (1995). Optimal designs for mixture amount experiments. *Journal of Statistical Planning and Inference*, *48*, 241–246.

Karlin, S., & Studden, W. J. (1966). Optimal experimental designs. *The Annals of Mathematical Statistics*, *37*, 783–815.

Kasatkin, O. G. (1974). On the construction of D-optimal design on a simplex (in Russian). Application of Mathematical methods for multi-component systems investigation. Metallurgia.

Kiefer, J., & Wolfowitz, J. (1960). The equivalence of two extremum problems. *Canadian Journal of Mathematics*, *12*, 363–366.

Kiefer, J. (1961). Optimum designs in regression problems II. *The Annals of Mathematical Statistics*, *32*, 298–325.

Kiefer, J. (1974). General equivalence theory for optimum designs (approximate theory). *The annals of Statistics*, *2*, 849–879.

Kiefer, J. (1975). Optimum design: Variation in structure and performance under change of criterion. *Biometrika*, *62*, 277–288.

Kiefer, J. (1978). Asymptotic approach to families of design problems. *Communications in Statistics-Theory and Methods*, *17*, 1347–1362.

Klein, T. (2004). Optimal designs for second-degree Kronecker model mixture experiments. *Journal of Statistical Planning and Inference*, *123*, 117–131.

Laake, P. (1975). On the optimal allocation of observations in experiments with mixtures. *Scandinavian Journal of Statistics*, *2*, 153–157.

Lambrakis, D. P. (1968). Experiments with mixtures: A generalization to the simplex-lattice design. Journal of the Royal Statistical Society. *Series B (Methodological)*, *30*, 123–136.

Lim, Y. B. (1990). D-optimal design for cubic polynomial regression on the q-simplex. *Journal of Statistical Planning and Inference, 25*, 141–152.
Liu, S., & Neudecker, H. (1995). A V-optimal design for Scheffé's polynomial model. *Statistics and Probability Letters, 23*, 253–258.
Pal, M., & Mandal, N. K. (2007). Optimum mixture design via equivalence theorem. *Journal of Combinatorics, Information & System Sciences, 32*, 107–126.
Pal, M., & Mandal, N. K. (2012). Optimum designs for estimation of parameters in a quadratic mixture-amount model. *Communications in Statistics-Theory and Methods, 41*, 665–673.
Piepel, G. F., & Cornell, A. J. (1985). Models for mixture experiments when the response depends on the total amount. *Technometrics, 27*, 219–227.
Prescott, P., Dean, A. M., Draper, N. R., & Lewis, S. M. (2002). Mixture experiments: ill conditioning and quadratic model formulation. *Technometrics, 44*, 260–268.
Prescott, P., & Draper, N. R. (2004). Mixture component-amount designs via projections, including orthogonally blocked designs. *Journal of Quality Technology, 36*, 413–431.
Scheffé, H. (1958). Experiments with mixtures. Journal of the Royal Statistical Society. *Series B (Methodological), 20*, 344–360.
Scheffé, H. (1963). Simplex-centroid design for experiments with mixtures. Journal of the Royal Statistical Society. *Series B (Methodological), 25*, 235–263.
Zhang, C. Q., Chan, L. Y., Guan, Y. N., Li, K. H., & Lau, T. S. (2005). Optimum designs for an additive quadratic mixture model involving the amount of mixture. *Statistica Sinica, 15*, 165–176.

Chapter 5
Optimal Mixture Designs for Estimation of Natural Parameters in Scheffé's Model Under Constrained Factor Space

Abstract Most of the studies on mixture experiments assume that the experimental region is the whole simplex. However, experimentation at the vertices of the simplex is generally not meaningful. One may, therefore, restrict the experiment to a subset of the simplex. In this chapter, an ellipsoid within the simplex is used as the experimental region, and Kiefer optimal designs are determined for both linear and quadratic models due to Scheffé.

Keywords Scheffé's linear and quadratic mixture models · Parameter estimation · Ellipsoidal factor space · Central composite design · Kiefer optimal designs

5.1 Introduction

The optimum designs in mixture experiment generally include the vertex points of the simplex as support points of the design, which are not mixture combinations in the true non-trivial sense. Practitioners who work with mixture experiments find such proposals rather illogical and absurd. They generally perform experiments excluding the vertex points. There are some suggestions (cf. Cornell 2002) which recommend taking the experimental region to be a subspace of the simplex that does not include the vertex points. This may be achieved by putting a constraint on the bounds of the components separately. In many situations, an ellipsoid within the simplex is used as the experimental region (cf. Cornell 2002, p. 109). Estimation of parameters and analysis can be found in the literature for experiments conducted within the ellipsoidal experimental region. Mandal et al. (2014) derived *Kiefer optimal* designs for parameter estimation in Scheffé (1958) models of degrees one and two when the experimental region is an ellipsoid within the simplex with center at the centroid. As is well known, most of the studies on mixture experiments assume the experimental region to be the whole simplex. In situations, where experimentation at the vertices of the simplex is not practicable, one may restrict the experiment to a subset of the simplex. Some authors (cf. Cornell 2002) suggested experimental region of the type

B. K. Sinha et al., *Optimal Mixture Experiments*, Lecture Notes in Statistics 1028,
DOI: 10.1007/978-81-322-1786-2_5, © Springer India 2014

$$\mathcal{X}_{CR} = \{\boldsymbol{x} = (x_1, x_2, \ldots, x_q) :, \quad x_i \geq 0, 1 \leq i \leq q, \sum_{i=1}^{q} x_i = 1,$$

$$(\boldsymbol{x} - \boldsymbol{x}_0)' H^{-2} (\boldsymbol{x} - \boldsymbol{x}_0) \leq 1\} \tag{5.1.1}$$

where $\boldsymbol{x}_0 = (1/q, 1/q, \ldots, 1/q)'$, $H = \text{diag}(h_{11}, h_{22}, \ldots, h_{qq})$. By varying h_{ii}s, one can control the experimental region to suit the specific situation. Mandal et al. (2014) investigated the problem of finding optimum designs for the estimation of parameters in mixture models of degrees one and two due to Scheffé (1958), when the experimental region is given by (5.1.1).

The true mean model, in the general form, is given by

$$\eta_{\boldsymbol{x}} = \boldsymbol{f}(\boldsymbol{x})' \boldsymbol{\beta}. \tag{5.1.2}$$

Let us write

$$\boldsymbol{z} = H^{-1} (\boldsymbol{x} - \boldsymbol{x}_0). \tag{5.1.3}$$

Then, because of the restrictions on x, given in (5.1.1), $\boldsymbol{z}$ must satisfy

$$\boldsymbol{z}' \boldsymbol{z} \leq 1. \tag{5.1.4}$$

$$\boldsymbol{z}' H \mathbf{1}_q = 0, \text{ or} \sum_i h_{ii} z_i = 0. \tag{5.1.5}$$

Let us assume that $H \propto I_q$, the identity matrix. Then, (5.1.5) simplifies to

$$\boldsymbol{z}' \mathbf{1}_q = 0. \tag{5.1.6}$$

Combining (5.1.4) and (5.1.6), the experimental region, in terms of $\boldsymbol{z}$, is therefore given by

$$\mathcal{X}_{\boldsymbol{z}} = \{\boldsymbol{z} : \boldsymbol{z}' \mathbf{1}_q = 0, \ \boldsymbol{z}' \boldsymbol{z} \leq 1\}. \tag{5.1.7}$$

Let Q be an orthogonal matrix defined as

$$Q = \begin{pmatrix} q^{-1/2} \mathbf{1}'_q \\ P^{(q-1) \times q} \end{pmatrix} \tag{5.1.8}$$

Clearly, P satisfies

$$P \mathbf{1}_q = 0, PP' = I_{q-1}, P'P = I_q - q^{-1} J_q. \tag{5.1.9}$$

Let us write

$$\begin{pmatrix} u \\ \boldsymbol{v} \end{pmatrix} = Qz$$

Then,

$$u = 0, \boldsymbol{v} = Pz, \tag{5.1.10}$$

and, because of (5.1.4) and (5.1.10), $\boldsymbol{v}$ satisfies

$$\boldsymbol{v}'\boldsymbol{v} \leq 1. \tag{5.1.11}$$

Thus, the model can be expressed in terms of a $(q-1)$-dimensional vector $\boldsymbol{v}$ satisfying (5.1.11). Now, the first degree model in the canonical form viz.

$$\eta_{\boldsymbol{x}} = \boldsymbol{\beta}'\boldsymbol{x} \tag{5.1.12}$$

can be expressed in terms of $\boldsymbol{v}$, using the transformations (5.1.3) and (5.1.8) as

$$\eta_z = \tau_0 + \boldsymbol{\tau}'\boldsymbol{v}, \tag{5.1.13}$$

where the relationship between the parameter vectors $\boldsymbol{\beta}$ and $\boldsymbol{\tau}^* = (\tau_0, \boldsymbol{\tau}')'$ in models (5.1.12) and (5.1.13) is given by

$$\boldsymbol{\tau}^* = R\boldsymbol{\beta}, \tag{5.1.14}$$

where R is a $p \times p$ matrix given by

$$R = \begin{pmatrix} 1 & 0 \\ 0 & P \end{pmatrix} \begin{pmatrix} \boldsymbol{x}_0' \\ H \end{pmatrix} \tag{5.1.15}$$

with $p = (q+1)q/2$.

The second degree response function in canonical form viz.

$$\eta_{\boldsymbol{x}} = \sum_i \beta_i x_i + \sum_{i<j} \beta_{ij} x_i x_j, \tag{5.1.16}$$

because of the constraint $\sum_i x_i = 1$, can be written as (cf. Pal and Mandal 2006)

$$\eta_{\boldsymbol{x}} = \boldsymbol{x}'B\boldsymbol{x} \tag{5.1.17}$$

where $B = (\beta_{ij}(1+\delta_{ij})/2)$, δ_{ij} being the Kronecker delta. As before, (5.1.17) can be expressed first as a function of z viz.

$$\begin{aligned}\eta_x &= (\boldsymbol{x}_0 + Hz)'B(\boldsymbol{x}_0 + Hz).\\ &= \theta_0 + z'\theta + z'Dz\\ &= \eta_z \text{ (say)},\end{aligned} \tag{5.1.18}$$

where θ_0, $\boldsymbol{\theta}$ and D are given by

$$\theta_0 = \boldsymbol{x}_0' B\boldsymbol{x}_0, \boldsymbol{\theta} = 2HB\boldsymbol{x}_0, D = HBH. \tag{5.1.19}$$

And now, (5.1.19) can be expressed as a function of $\boldsymbol{v}$ as follows:

$$\begin{aligned}\eta_x &= \tau_0 + \boldsymbol{v}'\boldsymbol{\tau} + \boldsymbol{v}'E\boldsymbol{v} \\ &= \eta_{\boldsymbol{v}}, \text{ say,}\end{aligned} \tag{5.1.20}$$

where

$$\tau_0 = \theta_0, \boldsymbol{\tau} = P\boldsymbol{\theta}, E = PDP', \boldsymbol{v} = Pz. \tag{5.1.21}$$

The problem is to find optimum designs for the estimation of the parameters of the models (5.1.13) and (5.1.20) with $\boldsymbol{v}$ satisfying (5.1.11).

5.2 Optimum Designs

In this subsection, optimum designs are derived first for the estimation of parameters in the models (5.1.13) and (5.1.20), in terms of $\boldsymbol{v}$, and subsequently, for the estimation of parameters in the original models viz. (5.1.2) and (5.1.17), in terms of the original mixing components $\boldsymbol{x}$ in the simplex.

The problem of determining optimum designs in terms of the variable $\boldsymbol{v}$ in the domain (5.1.11) is a standard one in the context of response surface design, and the results are well known. As mentioned earlier in Chaps. 2 and 4, Invariance structure, combined with the Loewner ordering, constitutes the Kiefer ordering, which provides an effective tool to attack optimal design problems of high dimension. The optimum design hence in terms of $\boldsymbol{x}$, can be obtained using (5.1.15) and (5.1.19)–(5.1.21). Since the optimum design in terms of $\boldsymbol{x}$ does not include the vertex points of the simplex, it is different from the standard optimum design obtained over the whole simplex.

5.2.1 First-order designs

One way to construct Kiefer optimal first-order design on the experimental domain (5.1.11) is to vary each of the $k = q - 1$ components on the two levels $\pm k^{-1/2}$ only.

The design that assigns uniform weight to each of the 2^k vertices of $[-k^{-1/2}, k^{-1/2}]^k$ is called the complete factorial design 2^k. The support size can be somewhat reduced by taking a fraction of the complete 2^k factorial, for which the associated model matrix X has orthogonal columns.

The following result helps to establish optimality of a design in the constrained region experiment:

Theorem 5.2.1 (cf. Pukelsheim 1993): *A first-order design $D(N \times k)$ with k factors is optimum in the sense of Kiefer ordering if $D'D \propto I_k$.*

As an example, consider the case of $q = 2$ components. Here, we have $k = 1$ variable, say v, in $[-1, +1]$. In order to estimate the two parameters of the transformed model:

$$\eta_v = \tau_0 + \tau v,$$

where $v \in [-1, +1]$, we need at least two support points. A design will be optimum in the sense of Kiefer-optimality criterion if it assigns equal mass at the two extremes viz. at $+1$ and -1.

From this, one can easily find the optimum design in the original factor space via z as follows.

For $q = 2$, we have $P = (2^{-1/2}, -2^{1/2})$. Then, the optimum design in the original factor space for different choices of H-matrix is given below:

(i) $H = 2^{-1/2} I_2, \boldsymbol{x} = \boldsymbol{x}_0 + Hz = (1, 0)$ for $v = 1$ and $(0, 1)$ for $v = -1$.
(ii) $H = 3^{-1/2} I_2, \boldsymbol{x} = \left(\frac{\sqrt{3}+\sqrt{2}}{2\sqrt{3}}, \frac{\sqrt{3}-\sqrt{2}}{2\sqrt{3}}\right)$ for $v = 1$ and $\left(\frac{\sqrt{3}-\sqrt{2}}{2\sqrt{3}}, \frac{\sqrt{3}+\sqrt{2}}{2\sqrt{3}}\right)$ for $v = -1$.
(iii)

$$H = 2^{-1} I_2, \boldsymbol{x} = \left(\frac{\sqrt{2}+1}{2\sqrt{2}}, \frac{\sqrt{2}-1}{2\sqrt{2}}\right) \text{ for } v = 1$$
$$\text{and } = \left(\frac{\sqrt{2}-1}{2\sqrt{2}}, \frac{\sqrt{2}+1}{2\sqrt{2}}\right) \text{ for } v = -1. \tag{5.2.1}$$

Theorem 5.2.1 establishes the Kiefer optimality of the designs obtained above for all choices of H of the form $H = hI_3$, where $h \le 2^{-1/2}$. From (i) above, we observe that the optimum design in the original factor space has support points at (1, 0) and (0, 1) with equal masses, which is known otherwise also. The Kiefer optimality of the same design has been established by Draper and Pukelsheim (1999) in their ingenious way.

A Kiefer optimum design with minimum number of support points corresponds to $k + 1$ vertices of the regular simplex with uniform mass $1/(k + 1)$. For $k = 2$, the three design points correspond to the three vertices of an equilateral triangle.

Similarly, for $k = 3$, the four design points correspond to the four vertices of an equilateral tetrahedron (cf. Pukelsheim 1993).

For $q = 3$, we get $k = q - 1 = 2$ and the 3 support points of the Kiefer optimum design are given by $(-\sqrt{3}/2, 1/2)'$, $(\sqrt{3}/2, 1/2)'$, and $(0, -1)'$. Let $H = 6^{-1/2} I_2$.

When $\boldsymbol{v} = (-\sqrt{3}/2, 1/2)'$, using the reverse transformations (5.1.10) and (5.1.3), where

$$P = \begin{pmatrix} -1/\sqrt{2} & 0 & 1/\sqrt{2} \\ 1/\sqrt{6} & -2/\sqrt{6} & 1/\sqrt{6} \end{pmatrix} \tag{5.2.2}$$

we get $\boldsymbol{x} = (1/6, 1/6, 2/3)'$. Similarly, corresponding to $\boldsymbol{v} = (\sqrt{3}/2, 1/2)'$ and $\boldsymbol{v} = (0, -1)'$, we get $\boldsymbol{x} = (2/3, 1/6, 1/6)'$ and $\boldsymbol{x} = (1/6, 2/3, 1/6)'$, respectively. This can also be viewed graphically:

The relationship between the parameter vectors $\boldsymbol{\beta}$ and $\boldsymbol{\tau}^* = (\tau_0, \boldsymbol{\tau}')'$ in models (5.1.12) and (5.1.14) is given by

$$\boldsymbol{\tau}^* = R\boldsymbol{\beta}, \tag{5.2.3}$$

where R is a $p \times p$ matrix given by

$$R = \begin{pmatrix} 1 & 0 \\ 0 & P \end{pmatrix} \begin{pmatrix} \boldsymbol{x}_0' \\ H \end{pmatrix} \tag{5.2.4}$$

with $p = (q+1)q/2$. Since the two matrices on right-hand side of (5.2.4) have rank q, using Sylvester's inequality, we have rank $I = q$. Hence, we can also represent (5.2.3) as

$$\boldsymbol{\beta} = R^{-1}\boldsymbol{\tau}. \tag{5.2.5}$$

For a given design ξ in (5.1.11), we can obtain estimate $\hat{\tau}$ of τ and hence $\hat{\boldsymbol{\beta}}$ of $\boldsymbol{\beta}$ using (5.2.5). Moreover, the dispersion matrix of the two estimates are related by

$$\text{Disp}(\hat{\boldsymbol{\beta}}) = R^{-1}\,\text{Disp}(\hat{\boldsymbol{\tau}})R'^{-1}.$$

Hence, the Loewner order domination of a design in terms of $\boldsymbol{\tau}$ also applies to $\boldsymbol{\beta}$. We can summarize the above findings in the following theorem.

Theorem 5.2.2 *The unique Kiefer optimal moment matrix is*

$$M = \begin{pmatrix} 1 & 0 \\ 0 & I_k \end{pmatrix}$$

The following theorem is also useful in finding a Kiefer optimal design.

Theorem 5.2.3 *The moment matrix remains invariant under rotation of the axes and retains the property of Kiefer optimality.*

Proof Let D be a design which is Loewner optimal.

For the first degree model, we have

$$W = (1, D).$$

Writing $W^* = (1/N^{1/2})W$, we get $M^* = W^{*\prime}W^* = I_{k+1}$.
Let D^* be a design obtained from D by a rotation of the axes. Then,

$$D^* = DU,$$

where $U(k \times k)$ is an orthogonal matrix.
Let, $W_U = (1, D^*) = WU^*$, where

$$U^* = \begin{pmatrix} 1 & 0 \\ 0 & U \end{pmatrix}$$

is a $(q \times q)$ orthogonal matrix.
Hence,

$$W_U = (1, D^*) = WU^*.$$

Then, it is easy to see that the moment matrices corresponding to W and W_U are identical.
Using Theorem 5.2.3, we have that any three points on the circumference of the circle in Fig. 5.1, which form an equilateral triangle, are Kiefer optimal.

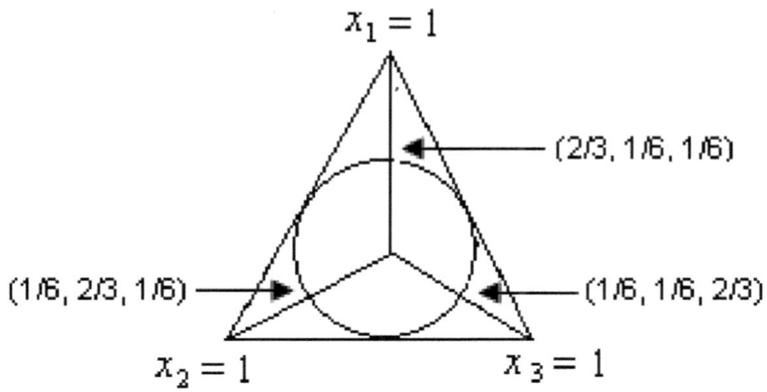

Fig. 5.1 Support points of the optimal first order design

5.2.2 Second-order designs

For the second degree model (5.1.17), one can find optimum designs in terms of the $q-1$ dimensional transformed vector $\boldsymbol{v}$. However, it is more convenient working with the Kronecker product representation of the model (cf. Pukelsheim 1993).

A Kronecker product representation:

Model (5.1.17) can be written as

$$\eta_x = \boldsymbol{x}' B \boldsymbol{x} = (\boldsymbol{x} \otimes \boldsymbol{x})' \boldsymbol{\beta}^*,$$

where $(\boldsymbol{x} \otimes \boldsymbol{x})$ is the symbolic direct product of the two vectors given by $(\boldsymbol{x} \otimes \boldsymbol{x})' = (x_1^2, x_1x_2, x_1x_3, \ldots, x_1x_q, x_2x_1, \ldots, x_2x_q, \ldots, x_q^2)$ and $\boldsymbol{\beta}^* = \overrightarrow{B} = (\beta_{11}^*, \beta_{12}^*, \ldots, \beta_{1q}^*, \beta_{21}^*, \ldots, \beta_{2q}^*, \ldots, \beta_{q1}^*, \ldots, \beta_{qq}^*)'$, where $\beta_{ii}^* = \beta_{ii}, \beta_{ij}^* = \beta_{ji}^* = \beta_{ij}/2$, for $i < j$.

Using the same transformations as before, we can express the response function η_x in terms of $\boldsymbol{v}$ via z, viz.:

$$\begin{aligned} \eta_v &= \tau_0 + \boldsymbol{v}'\boldsymbol{\tau}_1 + (\boldsymbol{v} \otimes \boldsymbol{v})'\boldsymbol{\tau}_2 \\ &= (1, \boldsymbol{v}', (\boldsymbol{v} \otimes \boldsymbol{v})')(\tau_0, \boldsymbol{\tau}_1', \boldsymbol{\tau}_2')' \\ &= \boldsymbol{h}'(\boldsymbol{v})\boldsymbol{\tau} \end{aligned} \tag{5.2.6}$$

where

$$\begin{aligned} \boldsymbol{h}'(\boldsymbol{v}) &= (1, \boldsymbol{v}', (\boldsymbol{v} \otimes \boldsymbol{v})'), \boldsymbol{\tau} = (\tau_0, \boldsymbol{\tau}_1', \boldsymbol{\tau}_2')', \\ \tau_0 &= \theta_0, \boldsymbol{\tau}_1 = P(\boldsymbol{\theta}_1 + \boldsymbol{\theta}_2), \boldsymbol{\tau}_2 = (P \times P)\boldsymbol{\theta}_3, \end{aligned} \tag{5.2.7}$$

and

$$\begin{aligned} \theta_0 &= (\boldsymbol{x}_0 \otimes \boldsymbol{x}_0)'\boldsymbol{\beta}^* \\ \boldsymbol{\theta}_1 &= (h/q) \begin{pmatrix} \mathbf{1}' & \mathbf{0}' & \mathbf{0}' & \ldots & \mathbf{0}' \\ \mathbf{0}' & \mathbf{1}' & \mathbf{0}' & & \mathbf{0}' \\ \ldots & \ldots & \ldots & \ldots & \ldots \\ \mathbf{0}' & \mathbf{0}' & \mathbf{0}' & & \mathbf{1}' \end{pmatrix} \\ \boldsymbol{\beta}^* &= (h/q)(\mathbf{1}' \otimes I_q)\boldsymbol{\beta}', \boldsymbol{\beta}_j^* = (\beta_{j1}^*, \beta_{j2}^*, \ldots, \beta_{jq}^*)' \\ \boldsymbol{\theta}_2 &= (h/q)(\mathbf{1}' \otimes I_q)P_{23}\boldsymbol{\beta}^*, \boldsymbol{\theta}_3 = h^2\boldsymbol{\beta}^* \end{aligned} \tag{5.2.8}$$

where $P_{23} = (\boldsymbol{e}_1, \boldsymbol{e}_3, \boldsymbol{e}_2, \boldsymbol{e}_4, \ldots, \boldsymbol{e}_q)$, $\boldsymbol{e}_i$ being a unit vector with the i-th element 1 and all other elements 0.

Combining above, the relation between $\boldsymbol{\beta}^*$ and $\boldsymbol{\tau}$ can be expressed as

$$\boldsymbol{\tau} = T\boldsymbol{\beta}^* \tag{5.2.9}$$

where

$$T = T_2 T_1,$$

$$T_1 = \begin{pmatrix} (1/q)^2(\mathbf{1}\otimes\mathbf{1})' \\ (h/q)(I\otimes\mathbf{1}')(I+P_{23} \\ h^2 I \end{pmatrix}, \quad T_2 = \begin{pmatrix} 1 & 0 & 0 \\ 0 & P & 0 \\ & & \\ 0 & 0 & P\otimes P \end{pmatrix} \tag{5.2.10}$$

Because of the linear relationship between $\boldsymbol{\tau}$ and $\boldsymbol{\beta}^*$, a design with Loewner order dominance in terms of $\boldsymbol{\tau}$ also carries over to $\boldsymbol{\beta}^*$. The optimality considerations in terms of $\boldsymbol{v}$ is a standard one in the context of response surface designs, and Mandal et al. (2013) exploited the standard results already available and then determined the optimum designs in the $\boldsymbol{x}$-space through reverse transformation.

Consider a central composite design (CCD) ξ^* in the domain $\{\boldsymbol{v} : \boldsymbol{v}'\boldsymbol{v} \le 1\}$, which is a mixture of three blocks of designs viz. cubes ξ_c, stars ξ_s and center points ξ_0 with suitable weights, where

ξ_c = regular 2^{k-r} fraction of the full factorial design (with levels $\pm 1/\sqrt{k}$), which is of resolution V (this ensures the estimation of all the parameters of the model. However, for $k < 5$, we have to take 2^k full factorial design),

ξ_s = set of star points of the form $(\pm 1, 0, 0, \ldots, 0), (0, \pm 1, 0, \ldots, 0), \ldots, (0, 0, \ldots, \pm 1)$,
$\xi_0 = \{\boldsymbol{v} \mid \boldsymbol{v}'\boldsymbol{v} = 0\}$.

Let ξ^* be given by

$$\xi^* = (1-\alpha)\xi_0 + \alpha\tilde{\xi}, 0 < \alpha < 1,$$

where

$$\tilde{\xi} = \frac{n_c\xi_c + n_s\xi_s}{n}, \quad n_c = k^2, n_s = 2^{k-r}, n = 2^{k-r}n_c + 2kn_s.$$

Such a design ξ^* is completely characterized by α.

For the model (5.2.6), the following result holds:

Theorem 5.2.4 *For the estimation of $\boldsymbol{\tau}$, the class of central composite designs (CCD) is complete in the sense that, given any design ξ, there is always a CCD which is better in terms of*

(i) Kiefer ordering
(ii) ϕ-optimality,

provided it is invariant with respect to orthogonal transformation.

(cf. Pukelsheim 1993).

Hence, the optimality of a CCD for the model (5.2.6) follows. Then, the optimum support points in terms of the original mixture components can be directly obtained from the transformations (5.1.3) and (5.1.10).

As we can express the parameters of the original model (5.1.17) as

$$\boldsymbol{\beta} = (\beta_{11}, \beta_{12}, \ldots, \beta_{1q}, \beta_{22}, \beta_{23}, \ldots, \beta_{qq})' = L\boldsymbol{\beta}^*,$$

for some matrix L of order $\frac{q(q+1)}{2} \times q^2$, it clearly follows that a design which has Loewner Order Dominance for $\boldsymbol{\beta}^*$ will also have Loewner Order Dominance for $\boldsymbol{\beta}$.

We, thus, have the following theorem.

Theorem 5.2.5 *The Kiefer optimal design for the estimation of parameters in a quadratic mixture model* (5.1.17) *having restricted experimental region* (5.1.1) *is obtained from a CCD, which is Kiefer optimal for the model* (5.1.2), *by using the transformations* (5.1.3) *and* (5.1.10).

It may be noted that the optimal value of α is determined from the optimality criterion used.

Example 5.2.1 Let us consider a three-component mixture. Here, we have $k = 2$. In order to estimate the parameters of the transformed model, we consider a CCD with the following support points:

(i) 4 star points: $(\pm 1, 0), (0, \pm 1)$
(ii) 2^2 factorial design points: $\frac{1}{\sqrt{2}}(-1, -1), \frac{1}{\sqrt{2}}(-1, 1), \frac{1}{\sqrt{2}}(1, -1), \frac{1}{\sqrt{2}}(1, 1)$
(iii) central point: $(0, 0)$.

Let $H = 6^{-1/2} I_3$.

Using the reverse transformation with

$$P = \begin{pmatrix} -1/\sqrt{2} & 0 & 1/\sqrt{2} \\ 1/\sqrt{6} & -2/\sqrt{6} & 1/\sqrt{6} \end{pmatrix}, \tag{5.2.11}$$

we get the nine support points in terms of $\boldsymbol{x}$ as:

1. $\left(\frac{1}{3} + \frac{1}{2\sqrt{3}}, \frac{1}{3}, \frac{1}{3} - \frac{1}{2\sqrt{3}}\right)$;
2. $\left(\frac{1}{3} - \frac{1}{2\sqrt{3}}, \frac{1}{3}, \frac{1}{3} + \frac{1}{2\sqrt{3}}\right)$;
3. $\left(\frac{1}{6}, \frac{2}{3}, \frac{1}{6}\right)$;
4. $\left(\frac{1}{2}, 0, \frac{1}{2}\right)$;
5. $\left(\frac{1}{3} + \frac{1}{2\sqrt{6}}, \frac{1}{3} + \frac{1}{3\sqrt{2}}, \frac{1}{3} - \frac{1}{2\sqrt{6}} - \frac{1}{3\sqrt{2}}\right)$;
6. $\left(\frac{1}{3} - \frac{1}{2\sqrt{6}} - \frac{1}{6\sqrt{2}}, \frac{1}{3} + \frac{1}{3\sqrt{2}}, \frac{1}{3} + \frac{1}{2\sqrt{6}} - \frac{1}{6\sqrt{2}}\right)$;
7. $\left(\frac{1}{3} + \frac{1}{2\sqrt{6}} + \frac{1}{6\sqrt{2}}, \frac{1}{3} - \frac{1}{3\sqrt{2}}, \frac{1}{3} - \frac{1}{2\sqrt{6}} + \frac{1}{6\sqrt{2}}\right)$;

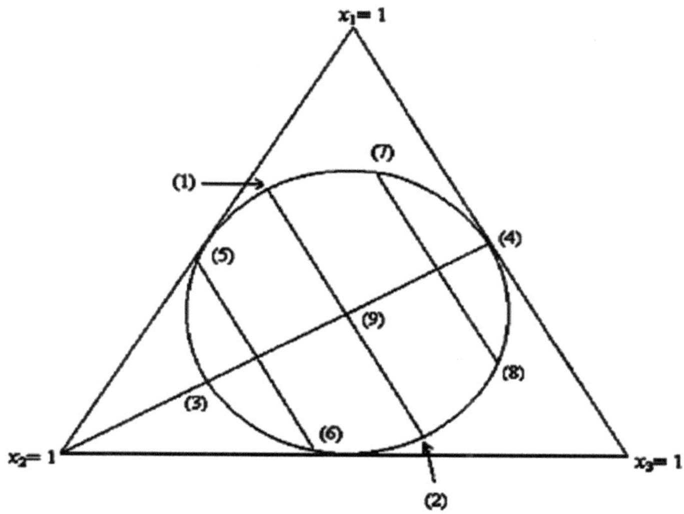

Fig. 5.2 Support points of optimal second-order design

8. $\left(\frac{1}{3} - \frac{1}{2\sqrt{6}} + \frac{1}{6\sqrt{2}}, \frac{1}{3} - \frac{1}{3\sqrt{2}}, \frac{1}{3} + \frac{1}{2\sqrt{6}} + \frac{1}{6\sqrt{2}}\right)$;
9. $\left(\frac{1}{3}, \frac{1}{3}, \frac{1}{3}\right)$.

Graphically, the support points are displayed in Fig. 5.2.

Remark 5.2.1 It can be easily shown that a design that is Kiefer optimal for any given P will remain so when P is replaced by $P^* = PT$, where T is a $q \times q$ permutation matrix, that is, a Kiefer optimal design will remain Kiefer optimal for any permutation of the mixing components.

References

Cornell, J. A. (2002). *Experiments with mixtures* (3rd ed.). New York: Wiley.

Draper, N. R., & Pukelsheim, F. (1999). Kiefer ordering of simplex designs for first and second degree mixture models. *Journal of Statistical Planning and Inference*, *79*, 325–348.

Mandal, N. K., Pal, M., Sinha, B. K., & Das, P. (2013). Optimum mixture designs in a restricted region. *Statistical Papers*. doi:10.1007/s00362-013-0568-0.

Mandal, N. K., Pal, M., & Sinha, B. K. (2014). Some finer aspects of the de la Garza Phenomenon: A study of exact designs in linear and quadratic regression models. To appear in *Statistics and Applications* (special volume in memory of Prof. M. N. Das)

Pal, M., & Mandal, N. K. (2006). Optimum designs for optimum mixtures. *Statistics & Probability Letters*, *76*, 1369–1379.

Pukelsheim, F. (1993). *Optimal design of experiments*. New York: Wiley.

Scheffé, H. (1958). Experiments with mixtures. *Journal of the Royal Statistical Society. Series B (Methodological)*, *20*, 344–360.

Chapter 6
Optimal Mixture Designs for Estimation of Natural Parameters in Other Mixture Models

Abstract In this chapter, we focus on finding optimum mixture designs for the estimation of natural parameters of models other than that of Scheffé viz., Becker's models, Darroch–Waller [D–W] model and Log-contrast model. It is also equally fascinating to note that so much has been done in these other mixture models as well. We mainly review the results that are already available and some new findings are presented.

Keywords Becker's models · Darroch–Waller model · Log-contrast model · Models with inverse terms · Estimation of parameters · Specific optimum designs

6.1 Introduction

In Chap. 4, we have discussed the optimality aspects of mixture designs for the estimation of natural parameters of Scheffé's models of different degrees. In Chap. 1, we have introduced different models in representing the mean response function. In this chapter, we will concentrate on finding optimum designs for the estimation of parameters of some models other than that of Scheffé.

6.2 Becker's Models

As was mentioned in Chap. 1, Becker (1968) proposed three distinct additive and homogeneous mixture models of degree one, involving synergism, as follows [see also Cornell and Gorman (1978)]:

B. K. Sinha et al., *Optimal Mixture Experiments*, Lecture Notes in Statistics 1028,
DOI: 10.1007/978-81-322-1786-2_6, © Springer India 2014

$$\eta(\boldsymbol{x}) = E(Y \mid \boldsymbol{x}) = \sum_i \beta_i x_i + \sum_{i<j} \beta_{ij} \frac{x_i x_j}{x_i + x_j} + \cdots$$
$$+ \sum_{i_1<i_2<\cdots<i_r} \beta_{i_1,i_2,\ldots,i_r} \frac{x_{i_1} x_{i_2} \ldots x_{i_r}}{(x_{i_1} + x_{i_2} + \cdots + x_{i_r})^{r-1}} \tag{6.2.1}$$

$$\eta(\boldsymbol{x}) = E(Y \mid \boldsymbol{x}) = \sum_i \beta_i x_i + \sum_{i<j} \beta_{ij} \min(x_i, x_j) + \cdots$$
$$+ \sum_{i_1<i_2<\ldots<i_r} \beta_{i_1,i_2,\ldots,i_r} \min(x_{i_1}, x_{i_2}, \ldots, x_{i_r}) \tag{6.2.2}$$

$$\eta(\boldsymbol{x}) = E(Y \mid \boldsymbol{x}) = \sum_i \beta_i x_i + \sum_{i<j} \beta_{ij} (x_i x_j)^{1/2} + \cdots$$
$$+ \sum_{i_1<i_2<\cdots<i_r} \beta_{i_1,i_2,\ldots,i_r} (x_{i_1} x_{i_2} \ldots x_{i_r})^{1/r}. \tag{6.2.3}$$

Liu and Neudecker (1997) introduced a *weighted simplex-centroid design* (WSCD) [afterward called weighted centroid design (WCD) by Draper and Pukelsheim (1999)] for a class of mixture models to which Becker's homogeneous response functions of degree one belong and attempted to find A-, D-, and I-optimal allocations of observations within this class of designs. They observed that for the model (6.2.2) with $r = 2$, the D-optimum design in WCD is *not* optimum in the entire class, when $q > 2$.

Pal et al. (2011a, b) restricted their investigations to $r = 2$ in finding D- and A-optimum designs for the models (6.2.1)–(6.2.3) in the entire class of competing designs with $q = 3$ components. For model (6.2.1) with $q = 2$, it is easy to observe that (2, 2) simplex design with equal mass at the support points is D-optimum. The same is true for models (6.2.2) and (6.2.3) as well. For A-optimality criterion, a design, which attaches weights α, α, and $1 - 2\alpha$ at the support points (1, 0), (0, 1), and (1/2, 1/2) respectively, is optimum in the entire class, where $\alpha = \sqrt{2}/(\sqrt{2}+1)$ for the models (6.2.2)–(6.2.3). Note that for $q = 2$, the model (6.2.1) reduces to a linear homogeneous model of Scheffé.

As mentioned above, Becker's models (6.2.1)–(6.2.3) consider all synergistic effects between components. However, in practice, it is quite possible to have a situation where only some of these effects exist. Pal et al. (2011a, b) considered three-component mixture models of the type (6.2.1)–(6.2.3) with synergism between some or all components and attempted to find optimum designs for the estimation of the model parameters, using D- and A-optimality criteria.

Consider model (6.2.1) with *exactly one* synergistic effect, say involving the first two components, Observe that apart from the synergistic effect β_{12}, other terms are linear in the three components x_1, x_2 and x_3. For such a model, it follows that the support points of the optimum design, for the estimation of parameters with respect to all the well-known optimality criteria, are the three vertices of the simplex. Since synergism is present between x_1 and x_2, and the models are invariant with respect to these two components, (1/2, 1/2, 0) is also likely to be a support point of the optimum design. In view of this, Pal et al. (2011a, b) confined to the subclass of designs $\mathcal{D}_1$ with support points as follows:

x_1	x_2	x_3	Mass
1	0	0	$\alpha_1/2$
0	1	0	$\alpha_1/2$
1/2	1/2	0	α_2
0	0	1	α_3

where $\alpha_i > 0, i = 1, 2, 3, \alpha_1 + \alpha_2 + \alpha_3 = 1$.

Since we are now in the framework of a saturated design, D-optimum design ξ_0 within $\mathcal{D}_1$ assigns weight $1/4$ to each of its support points. Using equivalence theorem of Kiefer (1974), the authors established the D-optimality of ξ_0 in the entire class of designs. This result also applies to the other two Becker's models displayed above. However, verification by way of application of Kiefer's Equivalence Theorem is highly non-trivial.

For the A-optimality criterion as well, the authors heuristically confined to $\mathcal{D}_1$. The optimum values of α_1, α_2, and α_3 came out as 0.4721, 0.4223, and 0.1056, respectively, for model (6.2.1); $2\sqrt{2}/(3 + 2\sqrt{2}), 2/(3 + 2\sqrt{2})$, and $1/(3 + 2\sqrt{2})$, respectively, for other two models (6.2.2)–(6.2.3). Then, the authors verified the optimality of these designs in the entire class using the equivalence theorem numerically.

Pal et al. (2011a, b) also considered a model where synergism is present between x_1 and x_2, as well as between x_1 and x_3. As before, because of invariance property of both the criteria with respect to the last two components, they restricted their search to the following subclass $\mathcal{D}_2$ of designs

x_2	x_3	Mass	x_1	Mass
1	0	1/2		
0	1	1/2	0	α_1
0	0	1	1	α_2
$1 - a$	0	1/2		
0	$1 - a$	1/2	a	α_3

where $\alpha_1 + \alpha_2 + \alpha_3 = 1$; α_1, α_2, and α_3 being the masses attached to $x_1 = 0, 1$, and a respectively, and $a \in (0, 1)$. The third column gives the masses for different (x_2, x_3) combinations when x_1 is given. It is noted that the mass at each of the five support points is to be taken as 1/5 since the design is saturated and we are looking for D-optimal design. Computations also provide optimum value of 'a' to be 1/2. Thus, the D-optimum design ξ_0 in $\mathcal{D}_2$ puts mass 1/5 at each of the extreme points (1, 0, 0), (0, 1, 0) and (0, 0, 1), and at each of the two mid-points of edges (1/2, 1/2, 0) and (1/2, 0, 1/2). The authors then examined the optimality or otherwise of ξ_0 in the entire class using equivalence theorem. Though the design ξ_0 satisfies the equivalence theorem for models (6.2.2)–(6.2.3) at all points in χ, but it fails to satisfy for model (6.2.1) for some points in the neighborhood of (1/3, 1/3, 1/3). This is a strikingly surprising result and it throws the optimality of the above design in question when we restrict to the entire domain χ. However, the absolute difference between the L.H.S. and R.H.S. of the equivalence theorem for these points is less than 0.2, which means that the absolute difference is less than 4 %. Hence, the design ξ_0 may be regarded as *nearly optimum* in the entire class of competing designs. Of course, starting with ξ_0, one may use some standard algorithm (cf. Silvey 1980) to find an even better design. We do not discuss this aspect any further.

For A-optimality criterion, for a design $\xi \in \mathcal{D}_2$, Pal et al. (2011a, b) derived the optimal values of α_1, α_2 and α_3 and these are given by $\alpha_1 = 0.2824$, $\alpha_2 = 0.1996, \alpha_3 = 0.5180$ which are incidentally the same for all the three models. Let us denote such a design by ξ_1. To verify the optimality or otherwise of ξ_1 in the entire class, they took recourse to the Equivalence Theorem. Numerical computations show that for $\xi^* = \xi_1$, equality in the condition of the equivalence theorem holds at all the support points for all the three models. The other condition of the equivalence theorem is satisfied at all other points for models (6.2.2)–(6.2.3) but is violated at many points in the domain χ, including the overall centroid point (1/3, 1/3, 1/3), and some points of the form $\left(\frac{1-b}{2}, \frac{1-b}{2}, b\right)$ for model (6.2.1). Hence, the design ξ_1 is A-optimum for models (6.2.2)–(6.2.3) but not for model (6.2.1). In view of the above observation, to find A-optimum design for model (6.2.1), Pal et al. (2011a) modified the class $\mathcal{D}_2$ to $\mathcal{D}_2^*$ involving six support points:

x_2	x_3	Mass	x_1	Mass
1	0	1/2		
0	1	1/2	0	α_1
0	0	1	1	α_2
$1-a$	0	1/2		
0	$1-a$	1/2	a	α_3
$\frac{1-b}{2}$	$\frac{1-b}{2}$	1	b	α_4

Numerical computation shows that, within the class of designs $\mathcal{D}_2^*$, the criterion function is minimized for the design ξ_0 having

$$a = 0.5155, b = 0.395$$
$$\alpha_1 = 0.28365, \alpha_2 = 0.16769, \alpha_3 = 0.42879, \alpha_4 = 0.11987.$$

It has been shown by the authors computationally that the condition of the equivalence theorem is satisfied and ξ_0 may indeed be considered as A-optimum in the entire class.

Pal et al. (2011a, b) also considered the case of three synergistic effects with the response function. Since for both D- and A-optimality criteria, the problem is invariant with respect to all the components, they proposed the following subclass of $(q, 2)$ simplex designs $\mathcal{D}_3$ with weights $\alpha/3$ at the three vertices and $(1-\alpha)/3$ at the three midpoints of the edges, where $0 < \alpha < 1$. As mentioned earlier, the number of support points of a design ξ in $\mathcal{D}_3$ being equal to the number of parameters to be estimated, for the D-optimality criterion, the masses at the support points must be all equal to 1/6. The authors observed that the above design is not D-optimum in the entire class for any of the three models which was also observed earlier by Liu and Neudecker (1997). The authors then modified the subclass of designs $\mathcal{D}_3$ to $\mathcal{D}_3^*$, by including (1/3, 1/3, 1/3) as a support point and obtained D-optimum design in this class. Such a design attaches weight α_1 at each of the axial points, α_2 at each of the midpoints of the edges and, finally, α_3 at the overall centroid point. The optimum values of α_i's came out as

$$\alpha_1 = \alpha_{10} = 0.16192, \alpha_2 = \alpha_{20} = 0.14557, \alpha_3 = \alpha_{30} = 0.07753, \qquad (6.2.4)$$

for all the three models. Moreover, by numerical investigation, they had enough reasons to claim that the design is D-optimum in the entire class.

For the A-optimality criterion, they obtained best design again in the subclass $\mathcal{D}_3$. However, as in D-optimality criterion, here again this design is not optimum in the entire class. Since the point (1/3, 1/3, 1/3) violated the condition of the equivalence theorem, they found best design in $\mathcal{D}_3^*$. The A-optimal design ξ_0, within $\mathcal{D}_3^*$, has the same weight (6.2.4) as for D-optimality for model (6.2.1) and

$$\alpha_1 = \alpha_{10} = 0.1423, \alpha_2 = \alpha_{20} = 0.1598, \alpha_3 = \alpha_{30} = 0.0313$$

for models (6.2.2)–(6.2.3). Finally, they verified the above design to be A-optimum in the entire class.

Becker's minimum polynomial of order r on the (unit) q-simplex including the minimum functions over all subsets of at most $r \leq q$ variables is considered by Hilgers (2000):

$$\eta_x = \sum_{i=1}^{q} \beta_i x_i + \sum_{r=2}^{q} \sum_{1 \leq i_1 < i_2 < \ldots < i_r \leq q} \beta_{i_1, i_2, \ldots, i_r} \min(x_{i_1}, x_{i_2}, \ldots, x_{i_r})$$

The model was applied in different kinds of scientific areas. D-optimal approximate designs for this model are shown to be supported on the barycenters. The minimum support design concentrated on the barycenters corresponding to the regression functions is optimal for $r = q$, whereas it fails to be optimal for $r < q$.

Heiligers and Hilgers (2003) established a close relation between admissible mixture and admissible mixture amount designs in additive and homogeneous models. This particularly allows us to obtain D-, A-, and V-optimal mixture amount from optimal mixture designs and vice versa. The authors presented some examples for Becker's and Scheffé's mixture models.

It may be mentioned that the problems of determining optimum designs for general r and q and for other optimality criteria are still open.

6.3 Darroch–Waller Model

The following additive quadratic model for the mean response function η_x was studied by Darroch and Waller (1985) for the case of $q = 3$:

$$\eta_{\mathrm{DW}2(\boldsymbol{x})} = \sum_{i=1}^{q} \alpha_i x_i + \sum_{i=1}^{q} \alpha_{ii} x_i (1 - x_i). \tag{6.3.1}$$

The model has been found to fit observed data well in many cases. For example, the model (6.3.1) fitted to the 6-component gasoline blending data for 'Motor' in Snee (1981) gave $R^2 = 0.990$ and R^2 (adjusted) $= 0.986$, and when fitted to the 4-comonent flare data in Snee (1973), gave $R^2 = 0.876$ and R^2 (adjusted) $= 0.751$. For $q = 3$, the model (6.3.1) and Scheffé's quadratic model are equivalent, but for $q = 2$ the parameters of the model (6.3.1) are not uniquely determined. For $q \geq 4$, (6.3.1) is a special case of Scheffé's quadratic model with the coefficients of Scheffé's model being governed by a system of linear constraints. The model (6.3.1) is additive in $x_1, x_2, \ldots, x_q$, and has fewer parameters than the Scheffé's model when $q \geq 4$.

Chan et al. (1995, 1998b) proved that the D-optimal saturated axial design has support points $x \leftrightarrow (1, 0, \ldots, 0)$ and $x \leftrightarrow (1 - (q - 1)1, 1, \ldots, 1)$, where $\delta_1 = 1/(q - 1)$ when $3 \leq q \leq 6$, and $\delta_1 = [(5q - 1) - (9q - 1)(q - 1)^{1/2}]/(4q^2)$ when $q \geq 7$.

For A-optimality criterion, by mimicking the arguments of Atwood (1969), Chan et al. (1998a) showed that only barycentres are possible support points for model (6.3.1). They obtained numerically the weights at the support points for different values of q and verified the A-optimality in the entire class using equivalence theorem. For completeness, we reproduce Table 6.1.

6.4 Log-Contrast Model

Different types of mixture models, such as polynomial and log-contrast models, have been developed to describe responses under mixture experiments. A linear log-contrast model introduced by Aitchison and Bacon-Shone (1984) is of the form

$$\eta(\boldsymbol{x}) = \beta_0 + \sum_{i=1}^{q} \beta_i \log x_i, \quad \sum_{i=1}^{q} \beta_i = 0. \tag{6.4.1}$$

The model (6.4.1), because of $\sum_{i=1}^{q} \beta_i = 0$, can also be written in terms of z_is as

$$\eta(\boldsymbol{x}) = \beta_0 + \sum_{i=1}^{q-1} \beta_i z_i, \, z_i = \log(x_i/x_q). \tag{6.4.2}$$

Table 6.1 *A*-optimal simplex-centroid designs for $\eta_{DW2(x)}$

q	$C(q,1)r_1$	$C(q,2)r_2$	$C(q,3)r_3$	$C(q,4)r_4$
3	0.3923	0.6077	0	–
4	0.4142	0.5858	0	0
5	0.3496	0	0.6504	0
⋮	⋮	⋮	⋮	⋮
21	0.4010	0	0.5990	0
22	0.3946	0	0.4687	0.1367
23	0.3881	0	0.3328	0.2791
24	0.3818	0	0.1974	0.4208
25	0.3769	0	0.0676	0.5565
26	0.3732	0	0	0.6268
⋮	⋮	⋮	⋮	⋮
$\to\infty$	0.3846	0	0	0.6154

The choice of the divisor x_q in z_i in Eq. (6.4.2) is arbitrary; indeed any one of $x_1, \ldots, x_{q-1}$ can be used as the divisor to produce different equivalent forms of $\eta(\boldsymbol{x})$. Writing $\boldsymbol{\beta} = (\beta_0, \beta_1, \ldots, \beta_{q-1})'$ and $\boldsymbol{z} = (1, z_1, \ldots, z_{q-1})'$, the form of Eq. (6.4.2) of $\eta(\boldsymbol{x})$ can be expressed as $\eta(\boldsymbol{x}) = \boldsymbol{\beta}'\boldsymbol{z}$. The form of Eq. (6.4.1) is more convenient for interpretation purposes, while Eq. (6.4.2) is more convenient for estimation of coefficients. The model $\eta(\boldsymbol{x})$ is suitable for describing a response that has drastic changes near the boundary of the simplex. This property of $\eta(\boldsymbol{x})$ and the usefulness of $\eta(\boldsymbol{x})$ in studying the additivity effect and inactiveness of mixture components on the response are studied in detail in Aitchison and Bacon-Shone (1984).

The model $\eta(\boldsymbol{x})$ in Eq. (6.4.2) is defined only in the interior of $\mathcal{X}$, because $\log x_i \to -\infty$ as $x_i \to 0^+$. For a given fixed constant $\delta \varepsilon (0, 1)$, define $a = -\log \delta (>0)$, and define the subset $\mathcal{X}(\delta)$ in the interior of $\mathcal{X}$ by

$$\mathcal{X}(\delta) = \{(x_1, \ldots, x_q) \ \varepsilon \ \mathcal{X}: \delta < x_i/x_j < 1/\delta \ (i, j = 1, \ldots, q)\}$$

in which the model $\eta(\boldsymbol{x})$ is well defined. Diagrams of $\mathcal{X}(\delta)$ for $q = 3, 4$ are shown in Chan (1988).

Chan (1988) showed that, if $q \geq 3$ is an odd integer, a *D*-optimal design for estimation of a full set of orthonormal beta-contrasts assigns weight $a/C(q, (q-1)/2)$ to each $\boldsymbol{x} \leftrightarrow (\delta, \ldots, \delta, 1, 1, \ldots, 1)((q-1)/2$ delta and $(q+1)/2$ ones) and weight $b/C(q, (q+1)/2))$ to each $\boldsymbol{x} \leftrightarrow (\delta, \ldots, \delta, 1, 1, \ldots, 1)((q+1)/2$ delta and $(q-1)/2$ ones) respectively, where $a, b \in (0, 1)$ are any numbers such that $a + b = 1$. This design is also *A*-optimal when $q > 3$ is an odd integer (Chan and Guan 2001).

If $q \geq 2$ is an even integer, (cf. Chan and Guan 2001) the D-optimal design assigns weight $1/C(q, q/2)$ to each $\boldsymbol{x} \leftrightarrow (\delta, \ldots, \delta, 1, 1, \ldots, 1)((q/2$ deltas and $q/2$ ones). This design in also A-optimal when q is an even integer.

As indicated by the authors, the advantage of a log-contrast model lies in the fact that as $z_i = \log(x_i/x_q)$ can be varied independently, the polynomial forms in z_is can be full in the sense of including all terms of appropriate degree, as against Scheffé's (1958) polynomial models in x_is which require the omission of certain terms to ensure identifiability. Moreover, this model can be used when the response changes drastically as $\boldsymbol{x}$ approaches the boundary of the domain $\boldsymbol{\chi}$.

Huang and Huang (2009a) presented an essentially complete class of designs under the Kiefer ordering for the linear log-contrast model. Based on the completeness result, ϕ_p-optimal designs for all p, $-\infty \leq p \leq 1$ including D- and A-optimal are obtained.

The quadratic log-contrast model proposed by Aitchison and Bacon-Shone (1984) is given by

$$\eta_{\boldsymbol{x}} = \beta_0 + \sum_{i=1}^{q-1} \beta_i \log(x_i/x_q) + \sum_{i=1}^{q-1}\sum_{j=1}^{q-1} \beta_{ij} \log(x_i/x_q) \log(x_j/x_q). \tag{6.4.3}$$

Chan (1992) discussed the D-optimal design for parameter estimation in (6.4.3) with experimental domain restricted to

$$\mathcal{X}^*(\delta) = \{(x_1, x_2, \ldots, x_q)' \in \text{rel.int.}\,\mathcal{X}\colon \delta \leq x_i/x_q \leq 1/\delta, i = 1, 2, \ldots, q-1\},$$
$$\delta \in (0, 1),$$

where relative interior of $\mathcal{X}$ is denoted by rel. int. $\mathcal{X}$. Note that the relative interior is defined in relation to a specified value of parameter $0 < \delta < 1$.

Huang and Huang (2009b) attempted to find D_s-optimal designs for *discriminating* between linear and quadratic log-contrast models by further restricting the experimental region to $\boldsymbol{\chi}(\delta)$ in the relative interior of $\mathcal{X}$ by fixing $\delta \in (0, 1)$ and defining

$$\mathcal{X}(\delta) = \{(x_1, x_2, \ldots, x_q)' \in \text{rel.int.}\,\mathcal{X}\colon \delta \leq x_i/x_j \leq 1/\delta,$$
$$\text{for all } i, j = 1, 2, \ldots, q\}. \tag{6.4.4}$$

Setting $t_i = -\log\left(\frac{x_i}{x_q}\right) / \log\delta$, model (6.4.3) can be written as

$$\eta(\boldsymbol{x}) \equiv \eta(\boldsymbol{t}) = \lambda_0 + \sum_{i=1}^{q-1} \lambda_i t_i + \sum_{i=1}^{q-1}\sum_{j=i}^{q-1} \lambda_{ij} t_i t_j$$
$$= f'(\boldsymbol{t})\boldsymbol{\lambda}, \tag{6.4.5}$$

where

$$f(\boldsymbol{t}) = (1, t_1, t_2, \ldots, t_{q-1}, t_1^2, t_2^2, \ldots, t_{q-1}^2, t_1t_2, t_1t_3, \ldots, t_{q-2}t_{q-1})',$$
$$\boldsymbol{\lambda} = (\lambda_0, \lambda_1, \ldots, \lambda_{q-1}, \lambda_{11}, \lambda_{22}, \ldots, \lambda_{q-1,q-1}, \lambda_{12}, \lambda_{13}, \ldots, \lambda_{q-2,q-1})',$$
$$\lambda_0 = \beta_0, \lambda_i = \beta_i(-\log\delta), \lambda_{ij} = \beta_{ij}(\log\delta)^2,$$

and the experimental domain is given by

$$\mathcal{F} = \{\boldsymbol{t} = (t_1, t_2, \ldots, t_{q-1})' \in [-1, 1]^{q-1}\colon t_i - t_j \in [-1, 1] \text{ for all } i, j = 1, 2, \ldots, q-1\}.$$

Let η_i be the uniform measure on the vertices of $\mathcal{F}$ with i coordinates equal to 1 or -1, denoted by

$$\eta_i = \begin{Bmatrix} \boldsymbol{t} \leftrightarrow (1, \ldots, 1, 0, \ldots, 0)', & \boldsymbol{t} \leftrightarrow (-1, \ldots, -1, 0, \ldots, 0)' \\ \dfrac{1}{2C(k-1, i)} & \dfrac{1}{2C(k-1, i)} \end{Bmatrix}$$

where $\boldsymbol{t} \leftrightarrow \boldsymbol{r}$ means $\boldsymbol{t} = P_n\boldsymbol{r}$ for some $P_n \in \text{Perm}(k-1)$, with $\text{Perm}(k-1)$ being the symmetric group consisting of all the $(k-1) \times (k-1)$ permutation matrices.

Using the technique of Lim and Studden (1988) for the invariant designs with respect to the group consisting of permutations and sign changes of the coordinates and the Equivalence Theorem to determine the D_s-optimal designs for the polynomial regression in q variables of degree n on the q-cube $[-1, 1]^q$, Huang and Huang (2009b) obtained optimum designs for the estimation of coefficients associated with second-order terms in (6.4.5). They showed that for a symmetric subspace of the finite dimensional simplex, there is a D_s-optimal design with the nice structure that puts a weight $1/2^{k-1}$ on the centroid of this subspace and the remaining weight is uniformly distributed on the vertices of the experimental domain.

Theorem 6.4.1 *For the second-degree model* (6.4.5) *on the experimental domain* $\mathcal{F}$ *with* $q \geq 3$, *a* D_s*-optimal for the quadratic parameters* λ_{ij}, $1 \leq i \leq j \leq k-1$, *is* $\eta^* = \frac{1}{2^{k-1}}\eta_0 + \sum_1^{k-1} \frac{C(k-1,i)}{2^{k-1}}\,\eta_i$.

Finally, Huang and Huang (2009b) discussed the D_s-efficiency of the D-optimal design for quadratic model and the design given by Aitchison and Bacon-Shone (1984).

6.5 Models with Inverse Terms

The following models with inverse terms (Draper and John 1977a) are used when the response changes drastically as x approaches the boundary of the factor space $\boldsymbol{\chi}$:

$$\eta_1(\boldsymbol{x}) = \eta_{q,1}(\boldsymbol{x}) + \sum_{i=1}^{q} \frac{\beta_{-i}}{x_i}$$

$$\eta_2(\boldsymbol{x}) = \eta_{q,2}(\boldsymbol{x}) + \sum_{i=1}^{q} \frac{\beta_{-i}}{x_i}$$

where $\eta_{q,1}(\boldsymbol{x})$, $\eta_{q,2}(\boldsymbol{x})$ are the Scheffé linear and quadratic polynomials. Cubic and quartic polynomials can also be extended to include inverse terms. These models are not defined on the boundary of $\boldsymbol{\chi}$ and the following design space may instead be used:

$$\boldsymbol{\chi}^{**}(\delta) = \{\boldsymbol{x} = (x_1, x_2, \ldots, x_q) \in \boldsymbol{\chi},\ x_i \geq \delta, \quad i = 1, 2, \ldots, q\}$$

when $q = 2$, for the model $\eta_1(\boldsymbol{x})$ in $\boldsymbol{\chi}^{**}(\delta)$, the D-optimum design assigns measure 1/4 to each of the points $\boldsymbol{x} \leftrightarrow (\delta, 1-\delta)$ and $\boldsymbol{x} \leftrightarrow (\delta_1, 1-\delta_1)$ where $\delta_1 = (1 - \frac{1}{2}(1 - \sqrt{1 + 6\delta - 6\delta^2 - 2\sqrt{4\delta + 5\delta^2 - 18\delta^3 + 9\delta^4}}))$. For model $\eta_2(\boldsymbol{x})$ in $\boldsymbol{\chi}^{**}(\delta)$, the D-optimum design assigns measure 1/5 to each of the points $\boldsymbol{x} \leftrightarrow (\delta, 1-\delta)$, $\boldsymbol{x} \leftrightarrow (\delta_1, 1-\delta_1)$ and $\boldsymbol{x} = (1/2, 1/2)$, where $\delta_1 = [1 - 1 + 2(\delta - \delta^2)/3 - 2(12\delta - 11\delta^2 - 2\delta^3 + \delta^4/3)^{1/2^{1/2}}]/2$ (cf. Chan 2000). The analytic solution for general q is not yet known for $q \geq 3$. However, Draper and John (1977b) obtained numerical solutions for $q = 3, 4$ with $\delta = 0.05$.

References

Aitchison, J., & Bacon-Shone, J. (1984). Log contrast models for experiments with mixtures. *Biometrika*, *71*, 323–330.

Atwood, C. L. (1969). Optimal and efficient designs of experiments. *The Annals of Mathematical Statistics*, *40*, 1570–1602.

Becker, N. G. (1968). Models for the response of a mixture. Journal of the Royal Statistical Society. *Series B (Methodological)*, *30*, 349–358.

Chan, L. Y., Jiang, Y. C., & Guan, Y. N. (1995). D-optimal axial designs for an additive quadratic mixture model. In Proceedings of the Workshop on Quasi-Monte Carlo Methods and their Applications (pp. 166–172). Hong Kong: Hong Kong Baptist University, 11 Dec 1995.
Chan, L. Y. (1988). Optimal design for a linear log contrast model for experiments with mixtures. *Journal of Statistical Planning and Inference, 20*, 105–113.
Chan, L. Y. (1992). D-optimal design for a quadratic log contrast model for experiments with mixtures. *Communications in Statistics-Theory and Methods, 20*, 2909–2930.
Chan, L. Y., Meng, J. H., & Jiang, Y. C. (1998b). D-optimal axial designs for quadratic and cubic additive mixtures model. *Australian & New Zealand Journal of Statistics, 40*, 901–913.
Chan, L. Y., Guan, Y. N., & Zhang, C. (1998a). A-optimal designs for an additive quadratic mixtures model. *Statistica Sinica, 8*, 979–990.
Chan, L. Y. (2000). Optimal designs for experiments with mixtures: a survey. *Communications in Statistics-Theory and Methods, 29*, 2281–2312.
Chan, L. Y., & Guan, Y. N. (2001). A- and D-optimal designs for a log contrast model for experiments with mixtures. *Journal of Applied Statistics, 28*, 537–546.
Cornell, J. A., & Gorman, J. W. (1978). On the detection of an additive blending component in multi-component mixtures. *Biometrics, 34*, 251–263.
Darroch, J. N., & Waller, J. (1985). Additivity and interaction in three-component experiments with mixtures. *Biometrika, 72*, 153–163.
Draper, N. R., & St. John, R. C. (1977a). A mixture model with inverse terms. *Technometrics, 19*, 37–46.
Draper, N. R., & St. John, R. C. (1977b). Designs in three and four components for mixtures models with inverse terms. *Technometrics, 19*, 117–130.
Draper, N. R., & Pukelsheim, F. (1999). Kiefer ordering of simplex designs for first and second degree mixture models. *Journal of Statistical Planning and Inference, 79*, 325–348.
Heiligers, B., & Hilgers, R. D. (2003). A note on optimum mixture and mixture amount designs. *Statistica Sinica, 13*, 709–725.
Hilgers, R. D. (2000). D-optimal designs for Becker's minimum polynomial. *Statistics & Probability Letters, 49*, 175–179.
Huang, M. N. L., & Huang, M. K. (2009a). ϕ_p-optimal designs for a linear log contrast model for experiments with mixtures. *Metrika, 70*, 239–256.
Huang, M. K., & Huang, M. N. L. (2009b). D_s-optimality designs for quadratic log contrast model for experiments with mixtures. *Communications in Statistics-Theory and Methods, 38*, 1607–1621.
Kiefer, J. (1974). General equivalence theory for optimum designs (approximate theory). *The annals of Statistics, 2*, 849–879.
Lim, Y. B., & Studden, W. J. (1988). Efficient Ds-optimal designs for multivariate polynomial regression on the q-cube. *The Annals of Statistics, 16*, 1225–1240.
Liu, S., & Neudecker, H. (1997). Experiments with mixtures: optimal allocations for Becker's models. *Metrika, 45*, 53–66.
Pal, M., Mandal, N. K., & Aggarwal, M. L. (2011a). Optimum designs for parameter estimation in linear mixture model with synergistic effects. *Communications in Statistics-Theory and Methods, 40*, 2059–2073.
Pal, M., Mandal, N. K., Sinha, B. K., & Das, P. (2011b). Optimum designs in linear mixture models with synergistic effects. *Journal of Statistical-Theory Applied, 19*, 305–322.
Scheffé, H. (1958). Experiments with mixtures. Journal of the Royal Statistical Society. *Series B (Methodological), 20*, 344–360.
Silvey, S. D. (1980). *Optimal designs*. London: Chapman Hall.
Snee, R. D. (1973). Techniques for the analysis of mixture data. *Technometrics, 15*, 517–528.
Snee, R. D. (1981). Developing blending models for gasoline and other mixtures. *Technometrics, 23*, 119–120.

References

Chapter 7
Optimal Designs for Estimation of Optimum Mixture in Scheffé's Quadratic Model

Abstract This chapter examines the optimum designs for estimating the optimum mixing proportions in Scheffé's quadratic mixture model with respect to the A-optimality criterion. By optimum mixing proportion, we refer to the one that maximizes the mean response. Since the dispersion matrix of the estimate depends on the unknown model parameters, a *pseudo-Bayesian* approach is used in defining the optimality criterion. The optimum designs under this criterion have been obtained for two- and three-component mixtures. Further, using Kiefer's equivalence theorem, it has been shown that under invariant assumption on prior moments, the optimum design for a q-component mixture is a $(q, 2)$ simplex lattice design for $q = 3, 4$.

Keywords Scheffé's quadratic mixture model · Estimation of optimum mixing proportions · Trace criterion · Pseudo-Bayesian approach · Invariance · $(q\ ,\ 2)$-simplex lattice design · Kiefer's equivalence theorem

7.1 Introduction

As mentioned in Chap. 3, the response in a q-component mixture experiment depends on the proportions $x_1, \ldots, x_q$ of the mixing ingredients, and the experimental region is given by

$$\mathcal{X} = \{(x_1, x_2, \ldots, x_q) \mid x_i \geq 0, \quad i = 1, 2, \ldots, q; \quad \sum_i x_i = 1\}. \tag{7.1.1}$$

Scheffé (1958) introduced models of different degrees in canonical forms to represent the mean response function in a mixture experiment. The models have been reviewed in details in Chap. 3. The problem of finding optimum designs for estimation of the model parameters has been discussed in Chap. 4. However, the most important problem, no doubt, is to estimate the optimum proportions of mixing components that maximize the mean response.

B. K. Sinha et al., *Optimal Mixture Experiments*, Lecture Notes in Statistics 1028,
DOI: 10.1007/978-81-322-1786-2_7,

Box and Wilson (1951) first systematically considered the problem of determining the optimum factor combination in a quantitative multi-factor experiment. Later, a number of contributions along this line have been reported in the literature, see for example, Mandal (1978), Silvey (1980), Chatterjee and Mandal (1981), Mandal (1986), Mandal and Heiligers (1992), Fedorov and Müller (1997), Müller and Pazman (1998), Cheng et al. (2001), Melas et al. (2003) In the following sections we discuss optimality results relating to estimation of optimum mixture combination with respect to the A-optimality criterion.

7.2 Estimation of Optimum Mixing Proportions

Scheffé's quadratic mixture model gives the mean response as

$$\eta(\boldsymbol{x}) = \beta_0 + \sum_{i=1}^{q} \beta_i x_i + \sum_{i<j=1}^{q} \beta_{ij} x_i x_j. \tag{7.2.1}$$

Because of the constraint $\sum_{i=1}^{q} x_i = 1$, (7.2.1) can be equivalently represented as a standard mixture model:

$$\eta(\boldsymbol{x}) = \sum_{i=1}^{q} \beta_{ii} x_i^2 + \sum_{i<j=1}^{q} \beta_{ij} x_i x_j = x' B x, \tag{7.2.2}$$

where $\boldsymbol{x}' = (x_1, x_2, \ldots, x_q)$ and $B = ((1+\delta_{ij})\beta_{ij}/2)$, δ_{ij} being the kronecker delta with $\delta_{ij} = 1$ for $i = j$ and $\delta_{ij} = 0$ for $i \neq j$. We can further write (7.2.2) in the general linear model form, viz.

$$\eta(\boldsymbol{x}) = \boldsymbol{f}(\boldsymbol{x})' \boldsymbol{\beta},$$

where

$$f(\boldsymbol{x}) = (x_1^2, x_2^2, \ldots, x_q^2, x_1 x_2, \ldots, x_1 x_q, x_2 x_3, \ldots, x_2 x_q, \ldots, x_{q-1} x_q)',$$
$$\boldsymbol{\beta} = (\beta_{11}, \ldots, \beta_{qq}, \beta_{12}, \ldots, \beta_{1q}, \beta_{23}, \ldots, \beta_{2q}, \ldots, \beta_{q-1,q})'.$$

Suppose the response function (7.2.2) is concave with a finite maximum in the interior of the experimental region (7.1.1). Then, the optimum mixing combination maximizing the mean response is obtained as

$$\boldsymbol{\gamma} = \delta^{-1} B^{-1} \mathbf{1}_q, \tag{7.2.3}$$

where $\mathbf{1}_q$ is an unit vector and $\delta = (\mathbf{1_q}' B^{-1} \mathbf{1_q})$. It is clear that $\boldsymbol{\gamma}$ is a nonlinear function of the model parameters.

Given a continuous design ξ of the type (4.1.2), the unknown parameters β_{ij}s can be estimated by $\hat{\beta}_{ij}$s and an estimate of $\boldsymbol{\gamma}$ can be obtained as

$$\hat{\boldsymbol{\gamma}} = \hat{\delta}^{-1}\hat{B}^{-1}\mathbf{1}_q. \tag{7.2.4}$$

In large samples, adopting the δ-method, the dispersion matrix for $\hat{\boldsymbol{\gamma}}$ is given by

$$E[(\hat{\boldsymbol{\gamma}} - \boldsymbol{\gamma})(\hat{\boldsymbol{\gamma}} - \boldsymbol{\gamma})'] = A(\boldsymbol{\gamma})M^{-1}(\xi)A'(\boldsymbol{\gamma}), \tag{7.2.5}$$

where

$$A(\boldsymbol{\gamma}) = \left(\frac{\partial \boldsymbol{\gamma}}{\partial \beta_{11}}, \frac{\partial \boldsymbol{\gamma}}{\partial \beta_{22}}, \ldots, \frac{\partial \boldsymbol{\gamma}}{\partial \beta_{qq}}, \frac{\partial \boldsymbol{\gamma}}{\partial \beta_{12}}, \frac{\partial \boldsymbol{\gamma}}{\partial \beta_{13}}, \ldots, \frac{\partial \boldsymbol{\gamma}}{\partial \beta_{q-1,q}}\right),$$

and the information (moment) matrix of ξ for estimating $\boldsymbol{\beta}$ is

$$M(\xi) = \sum_i p_i \boldsymbol{f}(\boldsymbol{x}_i)\boldsymbol{f}(\boldsymbol{x}_i)'.$$

It can be shown that

$$A(\boldsymbol{\gamma}) = B^{-1}\Gamma(\boldsymbol{\gamma}), \tag{7.2.6}$$

where $\Gamma(\boldsymbol{\gamma})$ is given by

$$\Gamma(\boldsymbol{\gamma}) = \begin{pmatrix} 2(\gamma_1^2 - \gamma_1) & 2\gamma_2^2 \ldots\ldots\ldots\ldots 2\gamma_1\gamma_2 - \gamma_2 \ldots\ldots\ldots\ldots 2\gamma_{q-1}\gamma_q \\ 2\gamma_1^2 & 2(\gamma_2^2 - \gamma_2) \ldots\ldots\ldots 2\gamma_1\gamma_2 - \gamma_2 \ldots\ldots\ldots\ldots 2\gamma_{q-1}\gamma_q \\ \ldots\ldots\ldots & \ldots\ldots\ldots \qquad \ldots\ldots\ldots \\ 2\gamma_1^2 & 2\gamma_2^2 \ldots\ldots\ldots\ldots 2\gamma_1\gamma_2 \ldots\ldots\ldots\ldots 2\gamma_{q-1}\gamma_q - \gamma_q \\ 2\gamma_1^2 & 2\gamma_i^2 \ldots\ldots\ldots\ldots 2\gamma_1\gamma_2 \ldots\ldots\ldots\ldots 2\gamma_{q-1}\gamma_q - \gamma_q \end{pmatrix}$$

Expressing β_{ij}'s in terms of b^{ij}'s, where $B^{-1} = (b^{ij})$, it can be easily checked that in order that Eq. (7.2.3) holds, the elements of B^{-1} and $\boldsymbol{\gamma}$ must satisfy

$$\begin{aligned} b^{ij} &= \delta\gamma_i^2 + (q-1)d \quad \text{if} \quad i = j \\ &= \delta\gamma_i\gamma_j - d \quad \text{if} \quad i \neq j. \end{aligned} \tag{7.2.7}$$

where d is a constant given by $d = [\delta q^{q-2} \mid B \mid]^{-1/(q-1)}$.

The generic constant 'd' will frequently appear in this chapter.

Then, $A(\boldsymbol{\gamma})$ simplifies to

$$d\begin{pmatrix} -2(q-1)\gamma_1 & 2\gamma_2 & \cdots & 2\gamma_q & \gamma_1-(q-1)\gamma_2 & \cdots & \gamma_{q-1}+\gamma_q \\ 2\gamma_1 & -2(q-1)\gamma_2 & \cdots & 2\gamma_q & \gamma_2-(q-1)\gamma_1 & \cdots & \gamma_{q-1}+\gamma_q \\ 2\gamma_1 & 2\gamma_2 & \cdots & 2\gamma_q & \gamma_1+\gamma_2 & \cdots & \gamma_{q-1}+\gamma_q \\ \cdots & \cdots & \cdots & \cdots & \cdots & \cdots & \cdots \\ 2\gamma_1 & 2\gamma_2 & \cdots & -2(q-1)\gamma_q & \gamma_1+\gamma_2 & \cdots & \gamma_{q-1}+\gamma_q \\ \cdots & \cdots & \cdots & \cdots & \cdots & \cdots & \cdots \\ 2\gamma_1 & 2\gamma_2 & \cdots & 2\gamma_q & \gamma_1+\gamma_2 & \cdots & \gamma_{q-1}-(q-1)\gamma_q \\ 2\gamma_1 & 2\gamma_2 & \cdots & 2\gamma_q & \gamma_1+\gamma_2 & \cdots & \gamma_q-(q-1)\gamma_{q-1} \end{pmatrix} \tag{7.2.8}$$

Any measure of accuracy in estimating γ will be a function of $E[(\hat{\boldsymbol{\gamma}}-\boldsymbol{\gamma})(\hat{\boldsymbol{\gamma}}-\boldsymbol{\gamma})']$, which is found to be dependent on the unknown model parameters. Since the mixture model, in its canonical form, is linear in the parameters, the information matrix $M(\xi)$ is independent of $\boldsymbol{\beta}$. Thus, $E[(\hat{\boldsymbol{\gamma}}-\boldsymbol{\gamma})(\hat{\boldsymbol{\gamma}}-\boldsymbol{\gamma})']$ depends on $\boldsymbol{\gamma}$ only through the elements of $A(\boldsymbol{\gamma})$. Of course, this is built upon the consideration that in the search for optimal design, one can 'disregard' the common multiplying factor 'd' in $A(\boldsymbol{\gamma})$. The problem of dependence of a measure on the unknown parameters can be tackled in several ways as follows:

(a) finding a locally optimum design by putting some specific values to the unknown parameters:
(b) finding optimum designs for different segments of the domain of unknown parameters;
(c) approaching sequentially; or
(d) adopting a Bayesian approach with some prior assumption on the distribution of the unknown parameters.

Another problem is the choice of the optimality criterion to determine the optimum design. In view of $\gamma'1_q = 1$, the matrix $E[(\hat{\boldsymbol{\gamma}}-\boldsymbol{\gamma})(\hat{\boldsymbol{\gamma}}-\boldsymbol{\gamma})']$ is singular, and hence, D-optimality criterion is ruled out. So we consider A-optimality criterion in this chapter.

Pal and Mandal (2006) adopted a pseudo-Bayesian approach to overcome the first problem. Since, without the factor d, the elements of the matrix $A(\boldsymbol{\gamma})$ are linear in the $\boldsymbol{\gamma}$-components, $E[(\hat{\boldsymbol{\gamma}}-\boldsymbol{\gamma})(\hat{\boldsymbol{\gamma}}-\boldsymbol{\gamma})']$ involves elements that are quadratic in $\gamma_i s$. Hence, they assumed a priori on the first two moments of the $\boldsymbol{\gamma}$-components, viz.

$$\mathcal{E}(\gamma_i^2) = v, i = 1, 2, \ldots, q; \quad \mathcal{E}(\gamma_i\gamma_j) = w, i \neq j = 1, 2, \ldots, q; \quad v > 0, w > 0. \tag{7.2.9}$$

Here, $v < 1/q$ since $1 = \mathcal{E}(\Sigma\gamma_i)^2 = qv + (q-1)w$ and $w > 0$. Moreover, this is useful in expressing $v(w)$ in terms of $w(v)$. It may further be noted that $1/q^2 < v < 1/q$.

The justification for taking the expectations of γ_i^2s to be equal across all i and also for taking the product moments to be all equal is that if nothing is known about the relative influence of the different components, there is no basis for assuming them to be unequal. Box and Hunter (1957), while introducing rotatability as a desirable property of a design, used a similar argument. If, however, some prior knowledge about the relative importance of different components are available, one can utilize it to have a prior different from (7.2.9). But, in that case, there is chance of losing the property of invariance among components, which might create difficulty in finding a closed form solution to the problem.

7.3 Optimum Mixture Designs Under Trace-Optimality Criterion

It is to be noted that since $\boldsymbol{\gamma}'\mathbf{1}_q = 1$, $E[(\hat{\boldsymbol{\gamma}}-\boldsymbol{\gamma})(\hat{\boldsymbol{\gamma}}-\boldsymbol{\gamma})']$ is singular. So, as a measure of comparison of different designs, Pal and Mandal (2006) used the criterion

$$\phi(\xi) = \text{Trace}\mathcal{E}\{E[(\hat{\boldsymbol{\gamma}}-\boldsymbol{\gamma})(\hat{\boldsymbol{\gamma}}-\boldsymbol{\gamma})']\},$$

where $\mathcal{E}$ stands for expectation with respect to the prior. From (7.2.5), $\phi(\xi)$ can be written as

$$\phi(\xi) = \text{Trace}\{M^{-1}(\xi)\mathcal{E}(A(\boldsymbol{\gamma})'A(\boldsymbol{\gamma})\}, \tag{7.3.1}$$

which is a linear optimality criterion (Fedorov 1972). A design will be said to be optimum if it minimizes $\phi(\xi)$.

The optimum designs have been explicitly obtained for $q = 2, 3$ and these are discussed below.

7.3.1 Case of Two Mixing Components

For the 2-component model, the moment matrix of a design ξ is given by

$$M(\xi) = \begin{pmatrix} \mu_{40} & \mu_{22} & \mu_{31} \\ & \mu_{04} & \mu_{13} \\ & & \mu_{22} \end{pmatrix},$$

where μ_{ij}s denote the product moments of order (i, j), and its inverse is expressed in the form

$$M^{-1}(\xi) = \begin{pmatrix} \mu^{40} & \mu^{22} & \mu^{31} \\ & \mu^{04} & \mu^{13} \\ & & \mu'^{22} \end{pmatrix}. \tag{7.3.2}$$

Whereas μ^{22} may result in a positive/negative/zero value, it necessarily follows that μ'^{22} is always positive. Consequently, the two are never equal.

The trace-optimality criterion function $\phi(\xi)$, given by (7.3.1), now simplifies to

$$\phi(\xi) = 2v(\mu^{40} + \mu^{04}) - 2(v - w)(\mu^{13} + \mu^{31}) + (v - w)\mu'^{22} - 4w\mu^{22}, \quad (7.3.3)$$

which is seen to be invariant with respect to the mixing components. Hence, using the result and terminology in Draper and Pukelsheim (1999), one can restrict to the class of Weighted Centroid Designs (WCDs) in finding the optimum design. It may be noted that WCDs are symmetric, invariant designs. Mixture designs in this subclass are denoted by η. In the expression (7.3.3), the parameters 'v' and 'w' are those in the prior distribution and are assumed to be known.

For the 2-component mixture, a WCD 'η' assigns mass $\alpha/2$ to each of the vertex points $(1, 0)'$ and $(0, 1)'$, and mass $(1 - \alpha)$ to the centroid $(1/2, 1/2)'$, $0 < \alpha < 1$. The information matrix based on the design η is given by

$$M(\eta) = (1/16)\begin{pmatrix} 1+7\alpha & 1-\alpha & 1-\alpha \\ & 1+7\alpha & 1-\alpha \\ & & 1-\alpha \end{pmatrix}. \quad (7.3.4)$$

Corresponding to the design η, (7.3.3) reduces to

$$\phi(\eta) = 2[s/\alpha + t/(1-\alpha)],$$

where s and t are both positive and are given by

$$s = 2(4v - 1) + 1/2, \quad t = 2(4v - 1).$$

Here, w is replaced by the corresponding expression in terms of v, given immediately after (7.2.9).

It is easy to see that the optimum choice of α is

$$\alpha_{opt} = s^{1/2}/(s^{1/2} + t^{1/2}) \quad (7.3.5)$$

and the minimum value of $\phi(\eta)$ is given by

$$\phi(\eta) = 2(s^{1/2} + t^{1/2})^2.$$

7.3.2 Case of Three Mixing Components

In this case the moment matrix of an arbitrary mixture design ξ has the form

$$M(\xi) = \begin{pmatrix} \mu_{400} & \mu_{220} & \mu_{202} & \mu_{310} & \mu_{301} & \mu_{211} \\ & \mu_{040} & \mu_{022} & \mu_{130} & \mu_{121} & \mu_{031} \\ & & \mu_{004} & \mu_{112} & \mu_{103} & \mu_{013}, \\ & & & \mu_{220} & \mu_{211} & \mu_{121} \\ & & & & \mu_{202} & \mu_{112} \\ & & & & & \mu_{022} \end{pmatrix}$$

where, as before, μ_{ijk} denotes the product moment of order (i, j, k).

Writing $M^{-1}(\xi)$ as

$$M^{-1}(\xi) = \begin{pmatrix} \mu^{400} & \mu^{220} & \mu^{202} & \mu^{310} & \mu^{301} & \mu^{211} \\ & \mu^{040} & \mu^{022} & \mu^{130} & \mu^{121} & \mu^{031} \\ & & \mu^{004} & \mu^{112} & \mu^{103} & \mu^{013} \\ & & & \mu'^{220} & \mu'^{211} & \mu'^{121} \\ & & & & \mu'^{202} & \mu'^{112} \\ & & & & & \mu'^{022} \end{pmatrix} \tag{7.3.6}$$

where μ'^{ijk} and μ^{ijk} need not be equal, the trace-optimality criterion function $\phi(\xi)$ reduces to

$$\begin{aligned} \phi(\xi) &= 24v\left(\mu^{400} + \mu^{040} + \mu^{004}\right) - 24w\left(\mu^{220} + \mu^{202} + \mu^{022}\right) \\ &\quad + 4(6w - 3v)\left(\mu^{310} + \mu^{301} + \mu^{013} + \mu^{031} + \mu^{130} + \mu^{103}\right) \\ &\quad - 12w\left(\mu^{211} + \mu^{121} + \mu^{112}\right) + (12v - 6w)\left(\mu'^{220} + \mu'^{202} + \mu'^{022}\right) \\ &\quad - 6v\left(\mu'^{211} + \mu'^{121} + \mu'^{112}\right). \end{aligned} \tag{7.3.7}$$

As in the case of two-component mixtures, (7.3.7) also enjoys the invariance property with respect to the components of the mixture. Thus, one can restrict to the class of WCDs, denoted by η, which have as their support points the three vertex points (η_1), the three midpoints of the edges (η_2) and the overall centroid point (η_3) of the simplex (7.1.1).

Let the masses assigned to η_1, η_2, and η_3 be given by α_1, α_2, and α_3, respectively with $\alpha_i \geq 0$ and $\Sigma\alpha_i = 1$. It is to be understood that all feasible solutions must ensure positive definiteness of the information matrix. In effect, the WCD η assigns mass

(i) $\alpha_1/3$ to each of the three vertex points $(1, 0, 0)'$, $(0, 1, 0)'$ and $(0, 0, 1)'$.
(ii) $\alpha_2/3$ to each midpoint of the edges $(1/2, 1/2, 0)'$, $(1/2, 0, 1/2)'$ and $(0, 1/2, 1/2)'$ and
(iii) α_3 to the overall centroid point $(1/3, 1/3, 1/3)'$.

Expressing α_3 in terms of α_1 and α_2, the moment matrix $M(\eta)$ of the design η comes out to be

Table 7.1 Optimum values of α_1 and α_2 and minimum $\phi(\eta)$ for some values of v

v	α_1	α_2	Min. $\phi(\eta)$
0.12	0.2565	0.7434	208.276
0.14	0.3066	0.6934	329.440
0.16	0.3285	0.6715	447.073
0.18	0.3410	0.6590	563.613
0.20	0.3492	0.6508	679.646
0.22	0.3551	0.6449	795.404
0.24	0.3589	0.6411	910.995
0.26	0.3620	0.6380	1026.479
0.28	0.3648	0.6352	1141.888
0.30	0.3670	0.6330	1257.244
0.32	0.3686	0.6314	1430.206
0.33	0.3693	0.6307	1430.206

$$M(\eta) = \begin{pmatrix} a & b & b & b & b & c \\ & a & b & b & c & b \\ & & a & c & b & b \\ & & & b & c & c \\ & & & & b & c \\ & & & & & b \end{pmatrix},$$

where $a = (1/81)(1 + 26\alpha_1 + 19\alpha_2/8), b = (1/81)(1 - \alpha_1 + 11\alpha_2/16)$, $c = (1/81)(1 - \alpha_1 - \alpha_2)$.

After some algebraic simplification, one gets

$$\begin{aligned}\phi(\eta) = 6[&(16\alpha_1 + \alpha_2)\{36v(2\alpha_1 + \alpha_2) - 3\alpha_2\} \\ &+ 4f(1 - 9v)(4\alpha_1 + \alpha_2)^2]/\alpha_1\alpha_2(8\alpha_1 + \alpha_2),\end{aligned} \quad (7.3.8)$$

where

$$f = -[32\alpha_1 + \alpha_2 - 2\alpha_1(8\alpha_1 - \alpha_2) - (4\alpha_1 + \alpha 2)^2]/[16\alpha_1 + \alpha_2 - (4\alpha_1 + \alpha_2)^2].$$

Note that, as before, w has been replaced by the corresponding expression in v while dealing with $q = 3$.

The problem, therefore, reduces to finding *feasible* α_1, α_2 (and hence α_3) so as to minimize (7.3.8) subject to the restrictions $\alpha_1, \alpha_2 \geq 0, \alpha_1 + \alpha_2 \leq 1$. However, it is rather difficult to find a closed form solution of the problem. Table 7.1 gives the optimum solutions and corresponding values of the criterion function for some selected values of v.

Remark 7.3.1 One can alternatively find an optimum design by minimizing

$$\int \text{Trace}\{\mathcal{E}((\hat{\boldsymbol{\gamma}} - \boldsymbol{\gamma})(\hat{\boldsymbol{\gamma}} - \boldsymbol{\gamma})')'\}d\boldsymbol{\gamma},$$

where the domain of integration, regarded as a function of $\boldsymbol{\gamma}$, may be the whole or a subspace of the simplex, depending on the available knowledge on $\boldsymbol{\gamma}$. Note that $\boldsymbol{\gamma}$ is otherwise akin to the mixing proportions in the experiment; vide (7.1.1). In case we deal with the whole of the simplex, it is not hard to show that the above criterion function is also invariant with respect to the components of the mixture and the optimum design is therefore a WCD.

Another way of finding an optimum design is by minimizing

$$\max[\text{Trace}\{\mathcal{E}((\hat{\boldsymbol{\gamma}} - \boldsymbol{\gamma})(\hat{\boldsymbol{\gamma}} - \boldsymbol{\gamma})')'\}],$$

where the maximum is taken with respect to $\boldsymbol{\gamma}$ in the simplex (7.1.1) or a subspace of it. This has been considered in Chap. 9. Such studies, in the context of general response surface design for the estimation of the extreme points, can be found in Mandal and Heiligers (1992), Müller (1995) and Müller and Pazman (1998).

Remark 7.3.2 With reference to the trace-minimization criterion, Table 7.1 reveals that for a three-component mixture experiment, the optimum design for estimation of the optimum mixture puts positive mass at the vertices and the midpoints of the edges, but zero mass at the overall centroid point. While discussing the estimation of parameters in the polynomial model of second degree, Kiefer (1961) also obtained D-optimal designs with support points at the vertices and the midpoint of the edges. Laake (1975) obtained the same support points in the second degree model using integrated variance criterion. From the table, it is further evident that the minimum trace increases with the value of v, which is in agreement with the fact that the more the variation in the information on $\boldsymbol{\gamma}$, the higher is the value of the criterion function. The same is also true for the case of two-component mixture. This is analytically verifiable.

Remark 7.3.3 Keeping in view the fact that the trace-optimal design assigns zero mass at the overall centroid (1/3, 1/3, 1/3), one could start with a subclass of designs having positive masses at the two other types of points viz., $\boldsymbol{x} \longleftrightarrow (1, 0, 0)$ and $\boldsymbol{x} \longleftrightarrow (1/2, 1/2, 0)$ i.e., one could confine to the points in a $(q, 2)$-simplex design and derive optimality results thereby. This is indeed possible and it turns out that the results can be generalized to the case of more than 3 components. These are discussed in the next section. The idea is to derive explicit forms of optimal designs within the subclass of $(q, 2)$-simplex designs and then to gainfully utilize Kiefer's equivalence theorem to assert the validity of the optimal nature of the design in the general class. It may be noted that $(q, 2)$-simplex design provides saturated design for estimation of the model parameters. A suitable reparametrization of the model (as is indicated in the next section) together with saturated nature of the $(q, 2)$-simplex mixture design enable us to go for an analytical derivation of the optimal design with respect to trace-optimality criterion.

7.4 Optimum Mixture Designs via Equivalence Theory

Kiefer (1974) established the general equivalence theorem that gives a necessary and sufficient condition for a design to be optimum in the entire class of competitive designs. The theorem is also helpful in getting an idea of the support points of the optimum design.

Consider the linear model given by (4.2.1). Let ϕ^* denote the optimality criterion and $\mathcal{M}$ be the class of all moment matrices M. Let $F_{\phi^*}\{M_1, M_2\}$ denote the Fréchet derivative of the criterion function ϕ^* at M_1 in the direction of M_2.

Theorem 7.4.1 (Equivalence Theorem) *For a design ξ, if ϕ^* be concave on $\mathcal{M}$ and differentiable at $M(\xi)$, then ξ is ϕ^*-optimal iff*

$$F_{\phi^*}\{M(\xi), \boldsymbol{f}(\boldsymbol{x})\boldsymbol{f}(\boldsymbol{x})'\} \leq 0 \tag{7.4.1}$$

for all $\boldsymbol{x}$ in the factor space.

Equality in (7.4.1) holds at the support points of ξ.

(Cf. Silvey 1980.)

In the problem considered in Sect. 7.2.1, where the criterion function ϕ, is given by (7.3.1), writing $\phi^* = -\phi$, and proceeding as in Silvey (1980), Theorem 7.4.1 reduces to the following:

Theorem 7.4.2 *A necessary and sufficient condition for a mixture design ξ to be trace-optimumis that*

$$\boldsymbol{f}(\boldsymbol{x})'M^{-1}(\xi)\{\mathcal{E}(A(\boldsymbol{\gamma})'A(\boldsymbol{\gamma}))\}M^{-1}(\xi)\boldsymbol{f}(\boldsymbol{x}) \leq Trace[M^{-1}(\xi)\mathcal{E}(A(\boldsymbol{\gamma})'A(\boldsymbol{\gamma}))] \tag{7.4.2}$$

holds for all $\boldsymbol{x} \in \mathcal{X}$.

Equality in (7.4.2) *holds at the support points of ξ.*

It is noted that the optimum designs for $q = 2, 3$, obtained in Sect. 7.2.1, belong to the class of $(q, 2)$-simplex lattice designs. However, for $q = 3$, the optimum designs for various combinations of prior moments were obtained numerically. In the present section, it shall be shown algebraically that for $q = 3$ and 4, the optimum design under the trace criterion (7.3.2) is a $(q, 2)$-simplex lattice design.

7.4.1 An Alternative Representation of Model (7.2.2)

For easy algebraic manipulation, the model (7.2.2) can be rewritten as

$$\eta(\boldsymbol{x}) = \sum_{i=1}^{q}\theta_{ii}x_i\left(x_i - \frac{1}{2}\right) + \sum_{i<j=1}^{q}\theta_{ij}x_ix_j, \tag{7.4.3}$$

where, writing $\boldsymbol{\theta} = (\theta_{11}, \theta_{22}, \ldots, \theta_{qq}, \theta_{12}, \ldots, \theta_{q-1,q})'$ one has $\boldsymbol{\theta} = P\boldsymbol{\beta}$, with

$$P = \begin{bmatrix} 2I_q & 0 \\ R & I_{C(q,2)} \end{bmatrix},$$

and R is a $C(q,2) \times q$ matrix given by

$$R = \begin{bmatrix} 1 & 1 & 0 & \ldots & 0 & 0 \\ 1 & 0 & 1 & \ldots & 0 & 0 \\ \cdots & \cdots & \cdots & \cdots & \cdots & \cdots \\ 1 & 0 & 0 & \ldots & 0 & 1 \\ 0 & 1 & 1 & \ldots & 0 & 0 \\ \cdots & \cdots & \cdots & \cdots & \cdots & \cdots \\ 0 & 1 & 0 & \ldots & 0 & 1 \\ \cdots & \cdots & \cdots & \cdots & \cdots & \cdots \\ 0 & 0 & 0 & \ldots & 1 & 1 \end{bmatrix}.$$

Hence, for a given design ξ, the estimate of $\boldsymbol{\theta}$ is obtained as $\hat{\boldsymbol{\theta}} = P\hat{\boldsymbol{\beta}}$, and

$$Disp(\hat{\boldsymbol{\theta}}) = P\,Disp(\hat{\boldsymbol{\beta}})P'. \tag{7.4.4}$$

Now, for a $(q,2)$-simplex design ξ with mass $p_1 = \frac{\alpha}{q}$ at each of the q vertices and mass $p_2 = \frac{1-\alpha}{C(q,2)}$ at each of the $C(q,2)$ midpoints of edges, where $0 \leq \alpha \leq 1$, the information matrix for $\boldsymbol{\theta}$ is obtained as

$$\Lambda(\xi, \boldsymbol{\theta}) = Diag\left(\overbrace{\frac{p_1}{4}, \ldots, \frac{p_1}{4}}^{q}, \overbrace{\frac{p_2}{16}, \ldots, \frac{p_2}{16}}^{C(q,2)}\right)'.$$

Hence, if $M(\xi, \boldsymbol{\beta})$ denotes the information matrix of ξ for estimating $\boldsymbol{\beta}$, from (7.4.4), it follows that

$$\begin{aligned} M^{-1}(\xi, \boldsymbol{\theta}) &= P^{-1}\Lambda^{-1}(\xi, \boldsymbol{\beta})P'^{-1} \\ &= \begin{bmatrix} \frac{1}{4}\Lambda_1^{-1} & -\frac{1}{4}\Lambda_1^{-1}R' \\ -\frac{1}{4}R\Lambda_1^{-1} & -\frac{1}{16}R\Lambda_1^{-1}R' + \Lambda_2^{-1} \end{bmatrix} \end{aligned} \tag{7.4.5}$$

where $\Lambda_1 = Diag(\frac{p_1}{4}, \ldots, \frac{p_1}{4})$, $\Lambda_2 = Diag(\frac{p_2}{16}, \ldots, \frac{p_2}{16})$.

The above representation of M^{-1} facilitates finding the optimum masses of the support points of a $(q,2)$-simplex design.

7.4.2 Case of Three Components

When $q = 3$, using (7.4.5), one obtains that for a (3, 2)-simplex design ξ with information matrix $M(\xi)$, the criterion function is given by

$$\phi(\xi) = 54d^2\left[\frac{11v - 10w}{\alpha} + \frac{16(2v - w)}{1 - \alpha}\right],$$

which is minimized at

$$\alpha = \alpha^* = \frac{\sqrt{(11v - 10w)}}{\sqrt{(11v - 10w)} + 4\sqrt{(2v - w)}},$$

and $\min \phi(\xi) = 54d^2[\sqrt{11v - 10w} + 4\sqrt{2v - w}]^2 = \phi^*$, say.

Remark 7.4.1 Essentially, if we set $\alpha_3 = 0$ and rewrite the expression (7.3.8) in terms of $\alpha(= \alpha_1$ in earlier notation), then optimization is possible and the above choice is the solution towards that.

Let ξ^* denote the design with $\alpha = \alpha^*$. Then one can easily check that

$$M(\xi^*)^{-1}\mathcal{E}(A(\boldsymbol{\gamma})'A(\boldsymbol{\gamma}))M(\xi^*)^{-1} = \begin{bmatrix} (a-b)I_3 + b\mathbf{1}_3\mathbf{1}_3' & C \\ & (g-h)I_3 + h\mathbf{1}_3\mathbf{1}_3' \end{bmatrix},$$

where

$$C = \begin{bmatrix} c & c & e \\ c & e & c \\ e & c & c \end{bmatrix},$$

$$a = d^2\frac{54(11v - 10w)}{\alpha^{*2}} = \phi^*, \quad b = d^2\frac{27(5v - 6w)}{\alpha^{*2}}$$

$$c = 27d^2\left[\frac{27v - 26w}{\alpha^{*2}} - \frac{16(5v - 6w)}{\alpha^*(1 - \alpha^*)}\right], e = 54d^2\left[-\frac{2(5v - 4w)}{\alpha^{*2}} - \frac{16(v - 2w)}{\alpha^*(1 - \alpha^*)}\right]$$

$$g = 54d^2\left[\frac{27v - 26w}{\alpha^{*2}} - \frac{32(5v - 6w)}{\alpha^*(1 - \alpha^*)} + \frac{256(2v - w)}{(1 - \alpha^*)^2}\right],$$

$$h = 27d^2\left[\frac{2(20v - 19w)}{\alpha^{*2}} + \frac{32(3v - 2w)}{\alpha^*(1 - \alpha^*)} - \frac{256v}{(1 - \alpha^*)^2}\right],$$

d being a constant independent of the design.

The condition (7.4.2) to be satisfied for optimality, therefore, simplifies to

$$a\sum_i x_i^4 + (2b + g)\sum_{i<j} x_i^2 x_j^2 + 2c\sum_{i \neq j} x_i^3 x_j + 2(e + h)\sum_{i \neq j \neq k} x_i^2 x_j x_k \leq \phi^*, \quad (7.4.6)$$

where $\phi^* = \phi(\xi^*)$

For the condition (7.4.6) to hold with equality sign at the support points of ξ^*, one must have

$$a = \phi^* \quad \text{and} \quad 2b + 4c + g = 14\phi^*,$$

which are true for ξ^*.

For any other point in the simplex, it is easy to see that the difference between the l.h.s. and r.h.s. of (7.4.6) is < 0 provided $2\phi^* - c > 0$, which is true since $c < 0$.

Thus, for $q = 3$, the Equivalence Theorem leads to the following theorem:

Theorem 7.4.3 *In a three-component quadratic mixture model, a* $(3, 2)$*-simplex lattice design with* $\alpha = \frac{\sqrt{11v - 10w}}{\sqrt{11v-10w+4\sqrt{2v-w}}}$ *is optimal in the whole class of competing designs, for given* v *and* w*, where* α *denotes the total mass at the extreme points.*

7.4.3 Case of Four Component Mixture

In the four-component mixture experiment, the criterion function ϕ, for a (4, 2)-simplex lattice design ξ with information matrix $M(\xi)$, is obtained as

$$\phi(\xi) = 2304d^2\left(\frac{v - w}{\alpha} + \frac{2(3v - w)}{1 - \alpha}\right),$$

which is minimized at

$$\alpha = \alpha^* = \frac{\sqrt{(v - w)}}{\sqrt{(v - w)} + \sqrt{2(3v - w)}},$$

and $\min \phi(\xi) = 2304d^2[\sqrt{v - w} + \sqrt{2(3v - w)}]^2 = \phi^*$, say.

Let ξ^* denote the design with $\alpha = \alpha^*$. Then one can easily check that

$$M(\xi^*)^{-1}\mathcal{E}(A(\boldsymbol{\gamma})'A(\boldsymbol{\gamma}))M(\xi^*)^{-1} = \begin{bmatrix} (a - b)I_3 + b\mathbf{1}_3\mathbf{1}_3' & C & D \\ & E & F \\ & & (e - f)I_3 + f\mathbf{1}_3\mathbf{1}_3' \end{bmatrix},$$

where

$$C = \begin{bmatrix} c & c & c \\ c & k & k \\ k & c & k \\ k & k & c \end{bmatrix}, D = \begin{bmatrix} k & k & k \\ c & c & k \\ k & c & k \\ k & c & c \end{bmatrix}, E = \begin{bmatrix} e & f & f \\ f & e & f \\ f & f & e \end{bmatrix}, F = \begin{bmatrix} f & f & g \\ f & g & f \\ g & f & f \end{bmatrix},$$

$$a = d^2 \frac{2304(v-w)}{\alpha^{*2}} = \phi^*, \quad b = d^2 \frac{256(v-w)}{\alpha^{*2}}$$
$$c = -d^2 \frac{256(v-w)}{\alpha^*}\left[\frac{10}{\alpha^*} + \frac{36}{1-\alpha^*}\right], k = d^2 \frac{512(v-w)}{\alpha^*}\left[-\frac{1}{\alpha^*} + \frac{6}{1-\alpha^*}\right]$$
$$e = 1024 d^2(v-w)\left[\frac{5}{\alpha^{*2}} + \frac{36}{\alpha^*(1-\alpha^*)} + \frac{72}{(1-\alpha^*)^2}\right],$$
$$f = 3072 d^2(v-w)\left[\frac{1}{\alpha^{*2}} + \frac{2}{\alpha^*(1-\alpha^*)} - \frac{12}{(1-\alpha^*)^2}\right].$$

The condition (7.4.2) to be satisfied for optimality of ξ^* simplifies to

$$a\sum_i x_i^4 + (2b+e)\sum_{i<j} x_i^2 x_j^2 + 2c\sum_{i \neq j} x_j^3 x_j + 2(k+f)\sum_{i \neq j \neq k} x_i^2 x_j x_k + 6g x_1 x_2 x_3 x_4 \leq \phi^*. \tag{7.4.7}$$

for all $\boldsymbol{x} \in \mathcal{X}$.

For equality to hold in (7.4.7) at the support points of ξ^*, one must have $a = \phi^*$ and $2b + 4c + e = \phi^*$, which are true. At all other points in $\mathcal{X}$ one can easily check that strict inequality holds provided $2\phi^* - c > 0$ and $4\phi^* - g > 0$, which also hold for ξ^*.

Thus, for $q = 4$, the Equivalence Theorem leads to the following:

Theorem 7.4.4 *In a four-component quadratic mixture model, a (4, 2)-simplex lattice design with* $\alpha = \frac{\sqrt{v-w}}{\sqrt{v-w}+\sqrt{2(3v-w)}}$ *is optimal in the whole class of competing designs for given prior moments v and w, where* α *denotes total mass at the extreme points.*

7.5 Optimum Mixture Designs with Unequal Apriori Moments

In Sects. 7.2.1 and 7.3.1, it is assumed that invariance exists among the second order prior moments of the $\boldsymbol{\gamma}$-components. In this section, a more general assumption on the prior moments has been made, viz.

$$\mathcal{E}(\gamma_i^2) = v_i, \ i = 1, 2, \ldots q \ \ \mathcal{E}(\gamma_i \gamma_j) = w_{ij}, \ j = 1, 2, \ldots, q; \ i < j. \tag{7.5.1}$$

Since $\sum_{i=1}^q \gamma_i = 1$, v_i, w_{ij}'s must satisfy

$$\sum_i v_i + 2\sum_{i<j} w_{ij} = 1.$$

The criterion function for the choice of design for estimating $\boldsymbol{\gamma}$ is, as before, $\phi(\xi)$, given by (7.3.1).

7.5.1 Case of Two Components

In the case of two components, since the design can be represented by points on a straight line of unit length in the two-dimensional space, the class of competing designs $\mathcal{D}$ can be substantially reduced using the following theorem:

Theorem 7.5.1 *Given any arbitrary design* $\xi \in \mathcal{D}$ *with information matrix* $M(\xi, \boldsymbol{\beta})$*, there exists a design* $\eta \in \mathcal{D}^* \subset \mathcal{D}$*, where* $\mathcal{D}^*$ *is the class of three-point designs with whole mass concentrated at the two extremes and a point in between, such that*

$$M(\eta, \boldsymbol{\beta}) \geq M(\xi, \boldsymbol{\beta}).$$

Proof Here the model is

$$E(Y \mid \boldsymbol{x}) = \zeta_x = \beta_{11}x_1^2 + \beta_{22}x_2^2 + \beta_{12}x_1x_2.$$

Since $x_1 + x_2 = 1$, we can rewrite the model as

$$\zeta_x = \beta_0^* + \beta_1^* x_1 + \beta_2^* x_1^2,$$

where $\boldsymbol{\beta}^* = (\beta_0^*, \beta_1^*, \beta_2^*)'$ and $\boldsymbol{\beta} = (\beta_{11}, \beta_{22}, \beta_{12})'$ are related by

$$\boldsymbol{\beta}^* = P\boldsymbol{\beta}, \tag{7.5.2}$$

with

$$P = \begin{pmatrix} 0 & 1 & 0 \\ 0 & -2 & 1 \\ 1 & 1 & -1 \end{pmatrix}.$$

Then,

$$Disp(\hat{\boldsymbol{\beta}}^*) = P Disp(\hat{\boldsymbol{\beta}})P'. \tag{7.5.3}$$

Let $\mathcal{D}_1$ be the set of single factor designs, based on x_1. Clearly, there is a one-to-one correspondence between $\mathcal{D}_1$ and $\mathcal{D}$. Further, one can write the moment matrix of any design $\xi_1 \in \mathcal{D}_1$ as $M(\xi_1, \boldsymbol{\beta}^*)$.

Now, let $\mathcal{D}_1^*$ be the set of single factor three-point designs, based on x_1. It is known (vide Liski et al. 2002) that for any arbitrary design $\xi_1 \in \mathcal{D}_1$, with information matrix $M(\xi_1, \boldsymbol{\beta}^*)$, one can find a design $\eta_1 \in \mathcal{D}_1^*$ with mass at 0, 1, and $a \in (0, 1)$ such that

$$M(\eta_1, \boldsymbol{\beta}^*) \geq M(\xi_1, \boldsymbol{\beta}^*)$$

in the Loewner Order Dominance sense.

Hence, from (7.5.2) and (7.5.3), it is clear that for any arbitrary two-component design $\xi \in \mathcal{D}$, there exists a three-point design $\eta \in \mathcal{D}^*$ such that $M(\eta, \boldsymbol{\beta}) \geq M(\xi, \boldsymbol{\beta})$.

This establishes the theorem.

From the theorem, it is clear that, in order to find the optimal design in the two factor case, one may restrict to the class of designs $\mathcal{D}^*$.

Let η denote a three-point design with masses α_1, α_2 and $1-\alpha_1-\alpha_2$, respectively, at the support points $(1, 0)$, $(0, 1)$ and $(a, 1-a)$, $a \in (0, 1)$. The information matrix for the design is then given by

$$M(\eta) = \begin{pmatrix} a_1 & b & c_1 \\ & a_2 & c_2 \\ & & b \end{pmatrix}, \tag{7.5.4}$$

where

$$\begin{aligned} a_1 &= \alpha_1 + a^4(1-\alpha_1-\alpha_2) \quad a_2 = \alpha_2 + (1-a)^4(1-\alpha_1-\alpha_2) \\ c_1 &= (1-\alpha_1-\alpha_2)a^3(1-a) \quad c_2 = (1-\alpha_1-\alpha_2)a(1-a)^3 \\ b &= (1-\alpha_1-\alpha_2)a^2(1-a)^2. \end{aligned}$$

In order to easily find the expression of the trace criterion, one can use an alternative representation of the response function following Pal and Mandal (2007):

$$\zeta_x = \theta_{11}x_1(x_1-a) + \theta_{22}x_2(x_2-(1-a)) + \theta_{12}x_1x_2, \tag{7.5.5}$$

where $\boldsymbol{\theta} = (\theta_{11}, \theta_{22}, \theta_{12})$ and $\boldsymbol{\beta}$ are related by

$$\boldsymbol{\beta} = L\boldsymbol{\theta},$$

with

$$L = \begin{pmatrix} 1-a & 0 & 0 \\ 0 & a & 0 \\ -a & -(1-a) & 1 \end{pmatrix}.$$

Then,

$$M(\xi, \boldsymbol{\theta}) = \begin{pmatrix} \alpha_1(1-a)^4 & 0 & 0 \\ & \alpha_2 a^4 & 0 \\ & & (1-\alpha)a^2(1-a)^2 \end{pmatrix},$$

where

$$\alpha = \alpha_1 + \alpha_2.$$

Hence,

$$M^{-1}(\xi, \boldsymbol{\beta}) = LM^{-1}(\xi, \boldsymbol{\theta})L' \tag{7.5.6}$$

Therefore,

$$\begin{aligned}\phi(\xi) &= \text{Trace}[LM^{-1}(\xi,\boldsymbol{\theta})L'\mathcal{E}(A'(\gamma)A(\gamma))]\\ &= \text{Trace}[M^{-1}(\xi,\boldsymbol{\theta})L'\mathcal{E}(A'(\gamma)A(\gamma))L]\\ &= \text{Trace}[M^{-1}(\xi,\boldsymbol{\theta})G],\ \text{say,}\end{aligned}$$

where

$$\begin{aligned}G &= ((g_{ij})) = L'\mathcal{E}(A'(\gamma)A(\gamma))L\\ g_{11} &= 8(1-a)^2v_1 + 2a^2(v_1+v_2-2w_{12}) - 8a(1-a)(w_{12}-v_1)\\ g_{22} &= 8a^2v_2 + 2(1-a)^2(v_1+v_2-2w_{12} - 8a(1-a)(w_{12}-v_2)\\ g_{33} &= 2(v_1+v_2-2w_{12})\\ g_{12} &= -2[2a^2(w_{12}-v_2) + 2(1-a)^2(w_{12}-v_1) - a(1-a)(v_1+v_2-6w_{12})]\\ g_{13} &= 2[2(1-a)(w_{12}-v_1) - a(v_1+v_2-2w_{12})]\\ g_{23} &= 2[2a(w_{12}-v_2) - (1-a)(v_1+v_2-2w_{12})].\end{aligned}$$

Thus, for given a,

$$\phi(\xi) = \frac{g_{11}}{\alpha_1(1-a)^4} + \frac{g_{22}}{\alpha_2 a^4} + \frac{g_{33}}{(1-\alpha)a^2(1-a)^2} \geq \left(\sum_i \sqrt{g_{ii}^*}\right)^2, \quad (7.5.7)$$

where

$$g_{11}^* = \frac{g_{11}}{(1-a)^4},\quad g_{22}^* = \frac{g_{22}}{a^4},\quad g_{33}^* = \frac{g_{33}}{a^2(1-a)^2},$$

and equality in (7.5.7) holds for

$$\alpha_i = \alpha_i^*(a) = \frac{\sqrt{g_{ii}^*}}{\sum_i \sqrt{g_{ii}^*}},\quad i = 1, 2. \quad (7.5.8)$$

Suppose a^* is the value of a minimizing $(\sum_i \sqrt{g_{ii}^*})^2$, $0 < a^* < 1$.

Then, the optimal design assigns masses $\alpha_1^*(a^*)$, $\alpha_2^*(a^*)$ and $1-\alpha_1^*(a^*)-\alpha_2^*(a^*)$, respectively, at the support points $(1, 0)$, $(0, 1)$ and $(a^*, 1-a^*)$.

7.5.2 Case of Three Component Mixture

Here, it is assumed that

$$v_1 = v_2,\quad w_{13} = w_{23}. \quad (7.5.9)$$

In Sect. 7.3.1, where there was invariance with respect to the prior moments, each component took three distinct values in the optimum design, two at the extremes and one in between. In the present case, the assumption (7.5.9) amounts to saying that we are treating the first two mixing components as 'exchangeable.' This, in its turn, presupposes that the 'optimum' mixing proportions also enjoy the same property. Thus, it leads to the *heuristic* argument that it may be enough to search for an optimum design in the hyperplane manifested by the property of exchangeability of the first two components. It turns out that in such a plane, the quadratic response surface function involving all the three-mixing components may be reduced to a quadratic in the third component only. Appealing to Liski et al. (2002), one can, therefore, adopt an initial design with x_3 taking the three values 0, 1, and some $a \in (0, 1)$.

Again, for any design ξ, using the expression (7.3.6) for $M^{-1}(\xi)$, the criterion function $\phi(\xi)$ comes out to be

$$\begin{aligned}\phi(\xi) &= 24v_1(\mu^{400} + \mu^{040}) + 24v_3\mu^{004} - 24w_{12}\mu^{220} - 24w_{13}(\mu^{202} + \mu^{022})\\ &+ 12(2w_{12} - v_1)(\mu^{310} + \mu^{130}) + 12(2w_{13} - v_1)(\mu^{301} + \mu^{031})\\ &+ 12(2w_{13} - v_3)(\mu^{013} + \mu^{103}) - 6(w_{12} + w_{13})(\mu^{211} + \mu^{121}) - 12w_{13}\mu^{112}\\ &+ 6(2v_1 - w_{12})\mu'^{220} + 6(v_1 + v_3 - w_{13})(\mu'^{202} + \mu'^{022})\\ &- 3(v_1 + w_{12} - w_{13})(\mu'^{211} + \mu'^{121}) - 6(v_3 + 2w_{13} - 2w_{12})\mu'^{112},\end{aligned}$$

which is invariant with respect to the first two components. Further, since $\phi(\xi)$ is convex with respect to the information matrix M, the optimum design will be invariant with respect to the first two components (Ref. Mandal et al. 2008a).

Hence, it seems reasonable to propose the following subclass of designs with support points as indicated below:

x_1	x_2	x_3	Mass
1	0	0	αW_1
0	1	0	αW_1
1/2	1/2	0	$(1-2\alpha)W_1$
0	0	1	W_2
$1-a$	0	a	$W_3/2$
0	$1-a$	a	$W_3/2$

where $0 < \alpha < 1/2, a \in (0, 1), W_i > 0, i = 1, 2, 3, W_1 + W_2 + W_3 = 1$. Let such a design be denoted by $\xi(a, \alpha, \boldsymbol{W})$. Then, after a little algebra, the information matrix for the design comes out to be

$$M(\xi) = D\Lambda D',$$

where D and Λ are, respectively, given by

$$D = d^2 \begin{pmatrix} \sqrt{\alpha} & 0 & b & 0 & \frac{(1-a)^2}{\sqrt{2}} & 0 \\ 0 & \sqrt{\alpha} & b & 0 & 0 & \frac{(1-a)^2}{\sqrt{2}} \\ 0 & 0 & b & 0 & 0 & 0 \\ 0 & 0 & 0 & 1 & \frac{a^2}{\sqrt{2}} & \frac{a^2}{\sqrt{2}} \\ 0 & 0 & 0 & 0 & \frac{a(1-a)}{\sqrt{2}} & 0 \\ 0 & 0 & 0 & 0 & 0 & \frac{a(1-a)}{\sqrt{2}} \end{pmatrix}, \quad \Lambda = Diag(W_1 I_3, W_2, W_3 I_2),$$

with b given by $\sqrt{\frac{1-2\alpha}{2^4}}$.

Hence,

$$\begin{aligned} \phi(\xi(a, \alpha, \boldsymbol{W})) &= \text{Trace } \Lambda^{-1}[D^{-1}\mathcal{E}(A'(\boldsymbol{\gamma})A(\boldsymbol{\gamma}))D'^{-1}] \\ &= \frac{g^*_{11}}{W_1} + \frac{g^*_{22}}{W_2} + \frac{g^*_{33}}{W_3}, \end{aligned}$$

where

$$\begin{aligned} &D^{-1}\mathcal{E}(A'(\boldsymbol{\gamma})A(\boldsymbol{\gamma}))D'^{-1} = ((g_{ij})), \\ &g^*_{11} = g_{11} + g_{22} + g_{33},\ g^*_{22} = g_{44},\ g^*_{33} = g_{55} + g_{66}, \\ &g_{11} = g_{22} = \tfrac{1}{\alpha}[24v_1 + 6(4v_1 - 5w_{12}) \\ &\qquad + 6(v_1 - w_{12} - 3w_{13})\tfrac{1-a}{a} + 6(v_1 + v_3 - w_{13})(\tfrac{1-a}{a})^2] = \tfrac{g^*}{2\alpha}, \text{ say}, \\ &g_{33} = 96\tfrac{(2v_1 - w_{12})}{1-2\alpha} = \tfrac{h^*}{1-2\alpha}, \text{ say}, \\ &g_{44} = 24v_3 + 24(v_3 - 2w_{13})\tfrac{\alpha}{1-\alpha} + 6(2v_1 + v_3 - 4w_{13} + 2w_{12})(\tfrac{1-a}{a})^2 \\ &g_{55} = g_{66} = 12\tfrac{(v_1 + v_3 - w_{13})}{a^2(1-a)^2}, \\ &g^* = 2 \times [24v_1 + 6(4v_1 - 5w_{12}) + 6(v_1 - w_{12} - 3w_{13})\tfrac{1-a}{a} \\ &\qquad + 6(v_1 + v_3 - w_{13})(\tfrac{1-a}{a})^2],\ h^* = 96(2v_1 - w_{12}). \end{aligned}$$

For given $a, \boldsymbol{W}, \phi(\xi(a, \alpha, \boldsymbol{W}))$ is minimized at $\alpha = \alpha_0 = \dfrac{\sqrt{g^*}}{2\sqrt{g^*} + \sqrt{2h^*}}$.

Then, at $\alpha = \alpha_0$,

$$\phi(\xi(a, \alpha_0, \boldsymbol{W})) = \phi(\xi(a, \boldsymbol{W})) = \frac{g^*_{11,0}}{W_1} + \frac{g^*_{22,0}}{W_2} + \frac{g^*_{33,0}}{W_3} \geq \left(\sum_i \sqrt{g^*_{ii,0}}\right)^2, \tag{7.5.10}$$

where $g^*_{ii,0} = g^*_{ii} \mid_{\alpha=\alpha_0}, i = 1, 2, 3$.

Equality holds in (7.5.10) when $W_i = W_i(a) = \dfrac{\sqrt{g^*_{ii,0}}}{\sum_i \sqrt{g^*_{ii,0}}}, i = 1, 2, 3$.

Hence, given a,

Table 7.2 Optimal designs for some combinations of $(v_1 = v_2, v_3, w_{12}, w_{13} = w_{23})$

Parameters				Optimal			
$v_1 = v_2$	v_3	w_{12}	$w_{13} = w_{23}$	a	α	W_1 W_2 W_3	Trace
0.2	0.2	0.0666	0.0666	0.5	0.2587	0.4497 0.1163 0.4473	679.645
0.2	0.1	0.15	0.05	0.4863	0.2514	0.4473 0.1077 0.4450	485.4117
0.2	0.1	0.12	0.065	0.4893	0.2509	0.4731 0.0949 0.4320	483.8437
0.2	0.1	0.10	0.075	0.4931	0.2505	0.4905 0.0854 0.4241	480.5998
0.15	0.2	0.12	0.065	0.4987	0.2501	0.4897 0.0863 0.4240	476.7452
0.1	0.2	0.065	0.1175	0.5094	0.2210	0.3818 0.1229 0.4953	285.6798

$$\phi(\xi(a, \alpha, \boldsymbol{W})) \geq \phi(\xi(a, \boldsymbol{W}(a))) = \left(\sum_i \sqrt{g^*_{ii,0}}\right)^2. \tag{7.5.11}$$

The optimal value of a is obtained so as to minimize the right-hand side of (7.5.11). However, algebraic deduction of optimal a is intractable. Table 7.2 gives the optimal values a, α and $\boldsymbol{W}$, for some combinations of $(v_1 = v_2, v_3, w_{12}, w_{13} = w_{23})$. It may be noted that the corresponding design is optimal only within the subclass

$$\mathcal{D}_0 = \{\xi(a, \alpha, \boldsymbol{W}); 0 \leq \alpha \leq 1, W_i \leq 0, i = 1, 2, 3, W_1 + W_2 + W_3 = 1\}. \tag{7.5.12}$$

However, with the help of Theorem 7.4.2, one can easily establish the optimality of the design $\xi_0 \equiv \xi(a^*, \boldsymbol{W}(a^*))$ within the entire class of competing designs.

Let $\phi(\xi_0) = \phi_0$ and $M^{-1}(\xi_0)\mathcal{E}(A'(\boldsymbol{\gamma})A(\boldsymbol{\gamma}))M^{-1}(\xi_0) = (b_{ij})$. Further, define

$$\begin{aligned} d(\xi_0, \boldsymbol{x}) &= \boldsymbol{f}(\boldsymbol{x})'M^{-1}(\xi_0)\mathcal{E}(A'(\boldsymbol{\gamma})A(\boldsymbol{\gamma}))M^{-1}(\xi_0)\boldsymbol{f}(\boldsymbol{x}) \\ &\quad - \text{Trace}[M^{-1}(\xi_0)\mathcal{E}(A'(\boldsymbol{\gamma})A(\boldsymbol{\gamma}))]. \end{aligned}$$

Since the matrix $M^{-1}(\xi_0)$ is symmetric and ξ_0 is necessarily invariant with respect to the first two components, we have

$$b_{13} = b_{23}, b_{14} = b_{24}, b_{15} = b_{26}, b_{16} = b_{25}, b_{35} = b_{36}, b_{45} = b_{46}, b_{55} = b_{66}.$$

Now, by virtue of Theorem 7.4.2, for the design ξ_0 to be optimal in the whole class, one must have $d(\xi_0, \boldsymbol{x}) = 0$ at the support points of ξ_0. For the condition to hold at the extreme points, it is essential that

$$b_{11} = b_{22} = b_{33} = \phi_0. \tag{7.5.13}$$

Under (7.5.13),

$$\begin{aligned} d(\xi_0, \boldsymbol{x}) = {} & x_1^2 x_2^2 (2b_{12} + b_{33} - 6b_{11}) + (x_1^3 x_2 + x_1 x_2^3)(2b_{13} - 4b_{11}) \\ & + x_3^2 (x_1^2 + x_2^2)(2b_{14} + b_{55} - 6b_{11}) + x_3 (x_1^3 + x_2^3)(2b_{15} - 4b_{11}) \\ & + x_3^3 (x_1 + x_2)(2b_{45} - 4b_{11}) + 2x_1 x_2 x_3 (x_1 + x_2)(b_{16} + b_{35} - 6b_{11}) \\ & + 2x_1 x_2 x_3^2 (b_{34} + b_{56} - 6b_{11}). \end{aligned} \tag{7.5.14}$$

So, to have $d(\xi_0, \boldsymbol{x}) = 0$ at the support points $(1/2, 1/2, 0)$, $(1 - a^*, 0, a^*)$ and $(0, 1 - a^*, a^*)$, b_{ij}s must satisfy

$$2b_{12} + b_{33} - 6b_{11} = -2(2b_{13} - 4b_{11}) \tag{7.5.15}$$

and

$$a^*(1-a^*)[(2b-4b_{11})a^{*2} + (2b_{15} - 4b_{11})(1-a^*)^2 + (2b_{14} + b_{55} - 6b_{11})a^*(1-a^*)] = 0. \tag{7.5.16}$$

Writing $A_1 = 2b_{45} - 4b_{11}$, $A_2 = 2b_{14} + b_{55} - 6b_{11}$, and using (7.5.15) and (7.5.16) in (7.5.14), $d(\xi_0, \boldsymbol{x})$ reduces to

$$\begin{aligned} d(\xi_0, \boldsymbol{x}) = {} & x_1 x_2 (x_1 - x_2)^2 (2b_{13} - b_{11}) + x_1 x_3 (1 - x_2)^2 \\ & \left[(A_1 + A_2 - A_3) \left(\frac{x_1}{1 - x_2} \right)^2 - (2A_2 - A_3 \left(\frac{x_1}{1 - x_2} \right) + A_2 \right] \\ & + x_2 x_3 (1 - x_1)^2 \left[(A_1 + A_2 - A_3) \left(\frac{x_2}{1 - x_1} \right)^2 - (2A_2 - A_3) \left(\frac{x_2}{1 - x_1} \right) + A_2 \right] \\ & + 2x_1 x_2 x_3 [x_3 (b_{34} + b_{56} b_{16} - b_{35}) + (b_{16} + b_{35} - 6b_{11})]. \end{aligned} \tag{7.5.17}$$

It is easy to check that (7.5.17) equals 0 at each support point of ξ_0.

Now, consider the quadratic form $h(y) = (A_1 + A_2 - A_3)y^2 - (2A_2 - A_3)y + A_2$, $0 \le y \le 1$. From (7.5.17), one gets $h(1-a^*) = 0$. Further, for $A_1 + A_2 - A_3 < 0$, and $A_3^2 = 4A_1 A_2$, $f(y)$ is a strictly concave function of y with maximum value 0 at $y = 1 - a^*$.

Thus, for any $\boldsymbol{x} \in \mathcal{X}, d(\xi_0, \boldsymbol{x}) \leq 0$ under the conditions

(i) $2b_{13} - 4b_{11} < 0$
(ii) $A_1 + A_2 - A_3 < 0$ and $A_3^2 = 4A_1A_2$
(iii) $b_{34} + b_{56} - b_{16} - b_{35} < 0,\ b_{16} + b_{35} - 6b_{11} < 0.$

The above discussion leads to a set of sufficient conditions for the design $\xi(a, \alpha, \boldsymbol{W})$ to be optimum within the entire class of competing designs:

Theorem 7.5.2 *A set of sufficient conditions for a mixture design $\xi(a, \alpha, \boldsymbol{W})$ with information matrix $M(\xi(a))$ and $M^{-1}(\xi(a))\mathcal{E}(A'(\boldsymbol{\gamma})A(\boldsymbol{\gamma}))M^{-1}(\xi(a)) = ((b_{ij}))$, and value of criterion function ϕ, to be optimal within the entire class of competitive designs is:*

$$\left.\begin{array}{l} (i)\ b_{11} = b_{22} = b_{33} = \phi \\ (ii)\ 2b_{13} - 4b_{11} < 0 \\ (iii)\ A_1 + A_2 - A_3 < 0 \text{ and } A_3^2 = 4A_1A_2 \\ (iv)\ a = \frac{2A_1 - A_3}{2(A_1 + A_2 - A_3)} \\ (v)\ b_{34} + b_{56} - b_{16} - b_{35} < 0,\ b_{16} + b_{35} - 6b_{11} < 0. \end{array}\right\} \quad (7.5.18)$$

where $A_1 = 3b_{45} - 4b_{11},\ A_2 = 2b_{15} - 4b_{11},\ A_3 = 2b_{14} + b_{55} - 6b_{11}.$

Extensive numerical computation shows that the optimum mixture design within the subclass $\mathcal{D}_0$, given by (7.5.12), satisfies the conditions (7.5.18). Thus, it appears that the optimum design in $\mathcal{D}_0$ is also optimum within the entire class of competing designs.

References

Box, G. E. P., & Wilson, K. B. (1951). On the experimental attainment of optimum conditions. Journal of the Royal Statistical Society. *Series B (Methodological)*, *13*, 1–38.

Box, G. E. P., & Hunter, J. S. (1957). Multi-factor experimental designs for exploring response surfaces. *The Annals of Mathematical Statistics*, *28*, 195–241.

Chatterjee, S. K., & Mandal, N. K. (1981). Response Surface Designs for estimating the optimal point. *Bulletin of the Calcutta Statistical Association*, *30*, 145–169.

Cheng, R. C. H., Melas, V. B., & Pepelyshev, A. N. (2001). Optimal designs for the evaluation of an extremum point. In V. B. Atkinson, B. Bogacka, & A. Zhigljavsky (Eds.), *Optimum design 2000* (pp. 15–24). Dordrecht: Kluwer Academic Publishers.

Draper, N. R., & Pukelsheim, F. (1999). Kiefer ordering of simplex designs for first and second degree mixture models. *Journal of Statistical Planning and Inference*, *79*, 325–348.

Fedorov, V. V. (1972). *Theory of optimal experiments*. New York: Academic.

Fedorov, V. V., & Müller, W. G. (1997). Another view on optimal design for estimating the point of extremum in quadratic regression. *Metrika*, *46*, 147–157.

Kiefer, J. (1961). Optimum designs in regression problems II. *The Annals of Mathematical Statistics*, *32*, 298–325.

Kiefer, J. (1974). General equivalence theory for optimum designs (approximate theory). *The annals of Statistics*, *2*, 849–879.

Laake, P. (1975). On the optimal allocation of observations in experiments with mixtures. *Scandinavian Journal of Statistics*, *2*, 153–157.

Liski, E. P., Mandal, N. K., Shah, K. R., & Sinha, B. K. (2002). *Topics in optimal design* (Vol. 163). Lecture notes in statistics New York: Springer.

Mandal, N. K. (1978). On estimation of the maximal point of a single factor quadratic response function. *Calcutta Statistical Association Bulletin*, *27*, 119–125.

Mandal, N. K. (1986). D-Optimum designs for estimating optimum points in a quantitative multiresponse experiment. *Bulletin of the Calcutta Statistical Association*, *35*, 37–49.

Mandal, N. K., & Heiligers, B. (1992). Minimax designs for estimating the optimum point in a quadratic response surface. *Journal of Statistical Planning and Inference*, *31*, 235–244.

Mandal, N. K., Pal, M., Sinha, B. K., & Das, P. (2008a). Optimum mixture designs: A Pseudo-Bayesian approach. Journal of the Indian Society of Agricultural Statistics, 62, 174–182. (special volume in memory of Prof. K. Kishen).

Melas, V. B., Pepelyshev, A., & Cheng, R. C. H. (2003). Designs for estimating an extremal point of quadratic regression models in a hyper-ball. *Metrika*, *58*, 193–208.

Müller, Ch H. (1995). Maximin efficient designs for estimating nonlinear aspects in linear models. *Journal of Statistical Planning and Inference*, *44*, 117–132.

Müller, Ch H, & Pazman, A. (1998). Applications of necessary and sufficient conditions for maximin efficient designs. *Metrika*, *48*, 1–19.

Pal, M., & Mandal, N. K. (2006). Optimum designs for optimum mixtures. *Statistics & Probability Letters*, *76*, 1369–1379.

Pal, M., & Mandal, N. K. (2007). Optimum mixture design via equivalence theorem. *Journal of Combinatorics, Information & System Sciences*, *32*, 107–126.

Scheffé, H. (1958). Experiments with mixtures. Journal of the Royal Statistical Society. *Series B (Methodological)*, *20*, 344–360.

Silvey, S. D. (1980). *Optimal designs*. London: Chapman Hall.

Chapter 8
More on Estimation of Optimum Mixture in Scheffé's Quadratic Model

Abstract Chapter 7 discusses the optimum designs for estimating the optimum mixture in Scheffé's quadratic mixture model, using the trace optimality criterion. In this chapter, we address the problem of finding optimum mixture designs under deficiency and minimax criteria. In most cases, Kiefer's equivalence theorem plays a key role in identifying the designs.

Keywords Scheffé's quadratic mixture model · Two- and three-component mixtures · Optimum mixing proportions · Deficiency criterion · Minimax criterion · Kiefer's equivalence theorem · Optimum designs

8.1 Introduction

As in Chap. 7, we consider the response η_x to be represented by Scheffé's quadratic mixture model in the form (7.2.2), which gives the optimum mixture as $\boldsymbol{x} = \boldsymbol{\gamma}$, given by (7.2.3). For comparing different designs to estimate $\boldsymbol{\gamma}$, Chatterjee and Mandal (1981) suggested the deficiency criterion, which identifies the optimum design as the one that minimizes the deficiency of the response at $\hat{\boldsymbol{\gamma}}$ from its optimum value. On the other hand, the minimax approach provides a tool to deal with the involvement of unknown parameters in any measure of accuracy in estimating $\boldsymbol{\gamma}$. This problem was overcome by using a pseudo-Bayesian approach in Chap. 7. In this chapter, we will discuss the above-mentioned optimality criteria for obtaining optimum designs to estimate $\boldsymbol{\gamma}$.

B. K. Sinha et al., *Optimal Mixture Experiments*, Lecture Notes in Statistics 1028, DOI: 10.1007/978-81-322-1786-2_8, © Springer India 2014

8.2 Optimum Mixture Designs Under Deficiency Criterion

Under model (7.2.2), and the assumptions made therein, the response is maximized at $\boldsymbol{x} = \boldsymbol{\gamma}$, given by (7.2.3). If $\hat{\boldsymbol{\gamma}}$ be an estimate of $\boldsymbol{\gamma}$, then $\eta_{\hat{\boldsymbol{\gamma}}}$ estimates the maximum response. Hence, the deficiency at $\hat{\boldsymbol{\gamma}}$ is measured by

$$\Psi\left(\boldsymbol{\gamma}, \hat{\boldsymbol{\gamma}}\right) = \eta_{\boldsymbol{\gamma}} - \eta_{\hat{\boldsymbol{\gamma}}} = \delta^{-1} - \hat{\boldsymbol{\gamma}}' B \hat{\boldsymbol{\gamma}},$$

where $\delta = \mathbf{1}_q' B^{-1} \mathbf{1}_q$.

For comparing different designs, Chatterjee and Mandal (1981) suggested the criterion of minimizing

$$E\Psi\left(\boldsymbol{\gamma}, \hat{\boldsymbol{\gamma}}\right) = \eta_{\boldsymbol{\gamma}} - E\left[\eta_{\hat{\boldsymbol{\gamma}}}\right] = \delta^{-1} - E\left[\hat{\boldsymbol{\gamma}}' B \hat{\boldsymbol{\gamma}}\right], \tag{8.2.1}$$

which is equivalent to maximizing $E[\hat{\boldsymbol{\gamma}}' B \hat{\boldsymbol{\gamma}}]$.

However, (8.2.1) depends on $\boldsymbol{\gamma}$ and the elements of B. This drawback can be easily resolved by adopting the pseudo-Bayesian approach discussed in Sect. 7.2.

Assuming prior knowledge about the moments of $\boldsymbol{\gamma}$ and B, the problem becomes that of maximizing

$$\mathcal{E} E\left[\hat{\boldsymbol{\gamma}}' B \hat{\boldsymbol{\gamma}}\right] = \mathcal{E}\left[\text{tr} B \mathcal{E}(\hat{\boldsymbol{\gamma}} \hat{\boldsymbol{\gamma}}')\right], \tag{8.2.2}$$

where $\mathcal{E}$ denotes the expectation with respect to suitably defined priors on $\boldsymbol{\gamma}$ and B and tr denotes trace. It is easily seen that maximizing (8.2.2) is equivalent to minimizing $\mathcal{E}[\text{tr} B^* E(\hat{\boldsymbol{\gamma}} \hat{\boldsymbol{\gamma}}')]$, where $B^* = -B$ is a positive definite matrix and $\mathcal{E}$ denotes expectation with respect to the priors on $\boldsymbol{\gamma}$ and B^*, equivalently stated.

Now,

$$B^* E\left(\hat{\boldsymbol{\gamma}} \hat{\boldsymbol{\gamma}}'\right) = B^* \boldsymbol{\gamma} \boldsymbol{\gamma}' + B^* V(\xi, \boldsymbol{\gamma}),$$

where

$$V(\xi, \boldsymbol{\gamma}) = A(\boldsymbol{\gamma}) M^{-1}(\xi) A'(\boldsymbol{\gamma}).$$

Since $B^* \boldsymbol{\gamma} \boldsymbol{\gamma}'$ is independent of the design, the problem reduces to that of minimizing

$$\begin{aligned} \phi(\xi) &= \text{tr}\mathcal{E}\left[B^* A(\boldsymbol{\gamma}) M^{-1}(\xi) A'(\boldsymbol{\gamma})\right] \\ &= \text{tr}\mathcal{E}_{\boldsymbol{\gamma}} \mathcal{E}_{B^*|\boldsymbol{\gamma}}\left[B^* A(\boldsymbol{\gamma}) M^{-1}(\xi) A'(\boldsymbol{\gamma})\right] \\ &= \text{tr}\left[M^{-1}(\xi) G\right], \end{aligned} \tag{8.2.3}$$

where $G = \mathcal{E}_{\boldsymbol{\gamma}} \mathcal{E}_{B^*|\boldsymbol{\gamma}}[A'(\boldsymbol{\gamma}) B^* A(\boldsymbol{\gamma})]$.

Optimum design is determined under the following assumptions:

(a) The prior of $\boldsymbol{\gamma}$ is given by (7.2.9).
(b) B^* has a prior distribution independent of $\boldsymbol{\gamma}$, and

$$\mathcal{E}_{B^*|\boldsymbol{\gamma}}[B^*] = \text{Diag}(a_1, a_2, \ldots, a_q) + b\mathbf{1_q}\mathbf{1_q}'. \tag{8.2.4}$$

Since $A'(\boldsymbol{\gamma})\mathbf{1_q} = \mathbf{0}$, G simplifies to $G = \mathcal{E}_{\boldsymbol{\gamma}}[A'(\boldsymbol{\gamma})\text{Diag}(a_1, a_2, \ldots, a_q)A(\boldsymbol{\gamma})]$.

8.2.1 Case of Two Mixing Components

Using the representation (7.3.2) for the inverse of the moment matrix of an arbitrary design ξ, the criterion function is obtained as

$$\begin{aligned}\phi(\xi) = (a_1 + a_2)\Big[&4v\left(\mu^{40} + \mu^{04}\right) - 8w\mu^{22} - 4(v - w)\left(\mu^{31} + \mu^{13}\right) \\ &+ 2(v - w)\mu^{22}\Big],\end{aligned}$$

which is clearly invariant with respect to the components of the mixture. Hence, by virtue of Draper and Pukelsheim (1999), search for optimum design may be restricted to the class of WCDs. Then, one can show that the optimum design is the same as that obtained for the two-component case in Sect. 7.3. It is worth noting that the optimum design is independent of the choices of a_1 and a_2. However, this is not true in general.

8.2.2 Case of Three Components

Here, again, writing the inverse of the moment matrix of any design ξ as (7.3.6), and $G = ((g_{ij}))$, the criterion function reduces to

$$\begin{aligned}\phi(\xi) = {}& g_{11}\mu^{400} + g_{22}\mu^{040} + g_{33}\mu^{004} + g_{44}\mu'^{220} + g_{55}\mu'^{202} + g_{66}\mu'^{022} + g_{66}\mu'^{022} \\ &+ 2\left(g_{12}\mu^{220} + g_{13}\mu^{202} + g_{24}\mu^{130} + g_{36}\mu^{013} + g_{23}\mu^{022}\right) \\ &+ 2\left(g_{14}\mu^{310} + g_{15}\mu^{301} + g_{35}\mu^{103} + g_{26}\mu^{031}\right) \\ &+ 2\left(g_{16}\mu^{211} + g_{25}\mu^{121} + g_{34}\mu^{112}\right) + 2\left(g_{45}\mu'^{211} + g_{46}\mu'^{121} + g_{56}\mu'^{122}\right),\end{aligned} \tag{8.2.5}$$

where

$$
\begin{aligned}
g_{11} &= 4v(4a_1 + a_2 + a_3), \\
g_{22} &= 4v(a_1 + 4a_2 + a_3), \\
g_{33} &= 4v(a_1 + a_2 + 4a_3), \\
g_{44} &= v(5a_1 + 5a_2 + 2a_3) - 2w(2a_1 + 2a_2 - a_3), \\
g_{55} &= v(5a_1 + 2a_2 + 5a_3) - 2w(2a_1 - a_2 + 2a_3), \\
g_{66} &= v(2a_1 + 5a_2 + 5a_3) - 2w(-a_1 + 2a_2 + 2a_3), \\
g_{12} &= 4w(-2a_1 - 2a_2 + a_3), \\
g_{13} &= 4w(-2a_1 + a_2 - 2a_3), \\
g_{23} &= 4w(a_1 - 2a_2 - 2a_3), \\
g_{14} &= -2v(2a_1 + 2a_2 - a_3) + 2w(4a_1 + a_2 + a_3), \\
g_{15} &= -2v(2a_1 - a_2 + 2a_3) + 2w(4a_1 + a_2 + a_3), \\
g_{24} &= -2v(2a_1 + 2a_2 - a_3) + 2w(a_1 + 4a_2 + a_3), \\
g_{25} &= -2w(a_1 + 4a_2 + a_3), \\
g_{26} &= -2v(-a_1 + 2a_2 + 2a_3) + 2w(a_1 + 4a_2 + a_3), \\
g_{34} &= -2w(a_1 + a_2 + 4a_3), \\
g_{35} &= -2v(2a_1 - a_2 + 2a_3) + 2w(a_1 + a_2 + 4a_3), \\
g_{36} &= -2v(-a_1 + 2a_2 + 2a_3) + 2w(a_1 + a_2 + 4a_3), \\
g_{16} &= -2w(4a_1 + a_2 + a_3), \\
g_{25} &= -2w(a_1 + 4a_2 + a_3), \\
g_{34} &= 2w(a_1 + a_2 + 4a_3), \\
g_{45} &= v(a_1 - 2a_2 - 2a_3), \\
g_{46} &= v(-2a_1 + a_2 - 2a_3), \\
g_{56} &= v(-2a_1 - 2a_2 + a_3). \qquad (8.2.6)
\end{aligned}
$$

It is evident from (8.2.6) that $\phi(\xi)$ is not invariant with respect to the components of the mixture when a_i's are unequal. As such, it is very difficult to find an optimum design for the problem when a_i's are all different. Mandal and Pal (2008) determined the optimum designs in two situations: (i) all a_i's are equal and (ii) any two of the a_i's are equal.

In case (i), it can be shown that $\phi(\xi) = a\phi^*(\xi)$, where a is the common value of a_i's and $\phi^*(\xi)$ is the trace criterion for a three-component mixture model. Then, by virtue of Theorem 7.4.3, the optimum design is a (3, 2)-simplex lattice design with the mass $\alpha = \dfrac{\sqrt{11v - 10w}}{\sqrt{11v - 10w} + 4\sqrt{v - w}}$ collectively assigned to the three extreme points, each with a share of $\alpha/3$ and mass $(1 - \alpha)/3$ to each of the three midpoints of the edges.

In case (ii), suppose it is assumed that $a_1 = a_2$, and a_3 is different. Under the assumption, the criterion function (8.2.5) is invariant with respect to the first two components of the mixture. Hence, if ξ_1 be a particular design with information matrix M_1 and one makes the exchange $x_1 \leftrightarrow x_2$ in the design to obtain a new design ξ_2 with information matrix M_2, then $\phi(\xi_1) = \phi(\xi_2)$. Further, since ϕ is convex and is invariant with respect to the above exchange, for the design $\overline{\xi} = \frac{1}{2}\xi_1 + \frac{1}{2}\xi_2$ whose information matrix is $\overline{M} = \frac{1}{2}(M_1 + M_2)$, one obtains

$$\phi(\overline{\xi}) = \text{tr}\left[\bar{M}^{-1}G\right] \le \text{tr}\left(\frac{1}{2}M_1^{-1}G + \frac{1}{2}M_2^{-1}G\right),$$

where equality holds if and only if $M_1 = M_2$.

Thus, we have the following result:

Result 8.2.1: *For a three-component mixture, when $a_1 = a_2$, the optimum design is necessarily invariant with respect to the first two components.*

In the case of $q = 2$, it has been observed that the optimum design has support points at $\{(1, 0), (0, 1), (\frac{1}{2}, \frac{1}{2})\}$, irrespective of the values of a_1 and a_2. Similarly, in the case of $q = 3$ with all a_i's equal, it has been seen that, irrespective of the common value of a_i's, the support points of the optimum design are at the vertices (0, 1, 0), (0, 1, 0), and (0, 0, 1) and at the midpoints of the edges $(\frac{1}{2}, \frac{1}{2}, 0)$, $(\frac{1}{2}, 0, \frac{1}{2})$, and $(0, \frac{1}{2}, \frac{1}{2})$ of the (3, 2)-simplex. In view of these observations, it appears that even when a_i's are not all equal, the support points of the optimum design should be confined to the same points. Again, when $a_1 = a_2$, in view of Result 8.2.1, the masses attached to the different support points will be symmetric with respect to the components x_1 and x_2. It, therefore, seems logical to restrict the search for optimum design in the following subclass of designs:

$$\begin{aligned}\mathcal{D} = \{\xi(\alpha_1, \alpha_3, \alpha_{12}, \alpha_{13});\ & \alpha_i \ge 0, i = 1, 3, \alpha_{12} \ge 0, \alpha_{13} \ge 0;\\ & 2(\alpha_1 + \alpha_{13}) + \alpha_3 + \alpha_{12} = 1\},\end{aligned}$$

where $\xi(\alpha_1, \alpha_3, \alpha_{12}, \alpha_{13})$ puts masses $(\alpha_1, \alpha_1, \alpha_3, \alpha_{12}, \alpha_{13}, \alpha_{13})$ at the support points $\{(0, 1, 0), (0, 1, 0), (0, 0, 1), (\frac{1}{2}, \frac{1}{2}, 0), (\frac{1}{2}, 0, \frac{1}{2}), (0, \frac{1}{2}, \frac{1}{2})\}$.

To find the optimal values of the masses, one can use the alternative representation (7.4.3) of the model, which simplifies algebraic computation substantially. For any $\xi \in \mathcal{D}$, the criterion function comes out to be

$$\begin{aligned}\phi(\xi) &= \frac{4(g_{11}^* + g_{22}^*)}{\alpha_1} + \frac{4g_{33}^*}{\alpha_3} + \frac{16g_{44}^*}{\alpha_{12}} + \frac{16(g_{55}^* + g_{66}^*)}{\alpha_{13}}\\ &\ge 4\left(\sum_{i=1}^{6}\sqrt{h_{ii}}\right)^2,\end{aligned} \quad (8.2.7)$$

where g^*_{ii} is equal to g_{ii}, given by (8.2.6), at $a_1 = a_2$, for $i = 1(1)6$ and

$$h_{ii} = g^*_{ii}, i = 1, 2, 3$$
$$= 4g^*_{ii}, i = 4, 5, 6.$$

Since $a_1 = a_2$, clearly $g^*_{11} = g^*_{22}, g^*_{55} = g^*_{66}$.
Hence, equality in (8.2.7) holds for

$$\alpha_i = \frac{\sqrt{h_{ii}}}{\sum_{i=1}^{6}\sqrt{h_{ii}}}, i = 1, 3; \quad \alpha_{12} = \frac{\sqrt{h_{44}}}{\sum_{i=1}^{6}\sqrt{h_{ii}}}; \quad \alpha_{13} = \frac{\sqrt{h_{55}}}{\sum_{i=1}^{6}\sqrt{h_{ii}}} \tag{8.2.8}$$

which give the optimal values of the masses.

Verification of the optimality of the above design in the entire class of designs can be done using Theorem 7.4.2. However, algebraic verification is intractable. Numerical computation with several points in $\mathcal{X}$ showed that equality in (8.3.1) holds at all the support points of the optimum design in $\mathcal{D}$, while for other points, strict inequality holds. Table 8.1 shows verification at three such points.

Remark 8.2.1 The optimum designs obtained for $q = 2, 3$ are $(q, 2)$-simplex lattice designs. It is likely that in the general case of q-component mixture, the optimum design will also be a $(q, 2)$-simplex design, even for arbitrary a_i's.

8.3 Optimum Mixture Designs Under Minimax Criterion

In this section, the problem of finding an optimum design for estimating the optimum mixture $\boldsymbol{\gamma}$ in model (7.2.2) has been considered using the minimax criterion. As has been mentioned in the beginning of Sect. 7.3, for any arbitrary design ξ with information matrix $M(\xi)$, the large sample dispersion matrix of the estimate $\hat{\boldsymbol{\gamma}}$ of $\boldsymbol{\gamma}$, given by (7.2.5), is singular. Hence, a suitable measure of accuracy of the design ξ can be taken to be

$$\varphi(\xi, \boldsymbol{\gamma}) = \text{Trace}\left[A(\boldsymbol{\gamma})M(\xi)^{-1}A(\boldsymbol{\gamma})'\right] \tag{8.3.1}$$

However, this measure depends on $\boldsymbol{\gamma}$. The minimax method overcomes this drawback by taking the supremum of $\varphi(\xi, \boldsymbol{\gamma})$ over $\boldsymbol{\gamma} \in \Gamma$, where $\Gamma = \{\boldsymbol{\gamma} \mid \gamma_i \geq 0, i = 1(1)q, \sum_{i=1}^{q}\gamma_i = 1\}$. The optimal design is then obtained by minimizing

$$\sup_{\boldsymbol{\gamma}\in\Gamma} \phi(\xi, \boldsymbol{\gamma}) = \sup_{\boldsymbol{\gamma}\in\Gamma} \text{Trace}\left[A(\boldsymbol{\gamma})M^{-1}(\xi)A'(\boldsymbol{\gamma})\right] \tag{8.3.2}$$

Table 8.1 The optimum design in the subclass of designs $\mathcal{D}$ and the value of $f(x)'M^{-1}GM^{-1}f(x)$ at three non-support points

Parameter values			Optimum masses at				Min. value of	$f(x)'M^{-1}GM^{-1}f(x)$ at		
			(1, 0, 0) and (0, 1, 0)	(0, 0, 1)	$(\frac{1}{2}, \frac{1}{2}, 0)$	$(\frac{1}{2}, 0, \frac{1}{2})$ and $(0, \frac{1}{2}, \frac{1}{2})$				
$a_1 = a_2$	a_3	v	α_1	α_3	α_{12}	α_{13}	Trace	$(\frac{1}{3}, \frac{1}{3}, \frac{1}{3})$	$(\frac{1}{3}, \frac{2}{3}, 0)$	$(\frac{1}{6}, \frac{1}{3}, \frac{1}{2})$
1	2	0.12	0.0801	0.0957	0.2693	0.2374	276.413	51.511	208.629	104.396
		0.20	0.1090	0.1303	0.2118	0.2200	903.661	456.312	773.084	470.535
		0.30	0.1152	0.1376	0.1977	0.2172	1822.89	1072.35	16.3.02	1.030.41
1	5	0.12	0.0730	0.1084	0.2968	0.2244	474.654	100.109	341.541	242.848
		0.20	0.0990	0.1469	0.2057	0.2247	1563.47	757.356	1261.31	881.470
		0.30	0.1044	0.1548	0.1832	0.2266	2876.40	1544.982	2388.61	1679.66
5	2	0.12	0.0908	0.742	0.2257	0.2592	828.814	154.240	669.721	211.944
		0.20	0.1236	0.1009	0.2227	0.2146	2710.04	1368.69	2467.52	1187.18
		0.30	0.1300	0.1061	0.2232	0.2054	5008.37	2909.98	4639.96	2461.05

Such minimax approach in the context of estimation of stationary point in a response surface problem has been considered earlier by Mandal and Heiligers (1992) and Cheng et al. (2001), among others.

In order to find the optimum design minimizing (8.3.2), one needs to study the properties of $\phi(\boldsymbol{\gamma}, M(\xi))$. It is noted that the experimental region $\mathcal{X}$ and the parameter space Γ are invariant with respect to permutation of components, i.e.,

$$R(\mathcal{X}) = \mathcal{X} \text{ and } R(\Gamma) = \Gamma,$$

for all $R \in \mathfrak{R}$, where $\mathfrak{R}$ is the class of all permutation matrices R of order $q \times q$.

Again, it is observed that as $\boldsymbol{x} \to R\boldsymbol{x}$, $f(\boldsymbol{x}) \to S_R f(\boldsymbol{x})$, where S_R, is a $p \times p$ orthogonal permutation matrix which depends on R and $p = C(q+1, 2)$. Under such a transformation, one has

$$M(\xi^R) = S_R M(\xi) S_R'.$$

Based on the above, we have the following properties of $\phi(\boldsymbol{\gamma}, M(\xi))$.

Property 8.3.1 *For fixed* $\boldsymbol{\gamma}$, $\phi(\boldsymbol{\gamma}, M(\xi))$ *is convex non-increasing in* $M(\xi)$ *with respect to Loewner partial ordering (LPO).*

Property 8.3.2 *Let* ξ^R *be the design obtained from* ξ *by virtue of the transformation* $\boldsymbol{x} \to R\boldsymbol{x}$. *Then, for any* $R \,\varepsilon\, \mathfrak{R}$, *we have*

$$\phi(\boldsymbol{\gamma}, M(\xi^R)) = \phi(R\boldsymbol{\gamma}, M(\xi)).$$

Property 8.3.3 *Let* $\phi_\Gamma(M(\xi)) = \sup_{\boldsymbol{\gamma} \in \Gamma} \phi(\boldsymbol{\gamma}, M(\xi))$. *Then,* $\phi_\Gamma(M(\xi))$ *is invariant with respect to* $R \in \mathfrak{R}$.

Theorem 8.3.1 follows from Properties 8.3.1 and 8.3.3:

Theorem 8.3.1 *If there exists a* Γ*-minimax design, then there exists a* Γ*-minimax design which is admissible and* S_R*-invariant.*

It is evident from the above theorem that in order to find the optimum design by minimax criterion, one can restrict to the class of invariant designs. By virtue of Draper and Pukelsheim (1999), one can further reduce the class by confining to WCDs.

In the case of two-component mixture, one can check that the optimum design assigns mass 0.2640 to each of the extreme points (1, 0) and (0, 1) and mass 0.4720 to the point $(\frac{1}{2}, \frac{1}{2})$.

In the case of mixture with three components, let η be a WCD with masses α_1, α_2, and α_3 at the extreme points, midpoints of edges and the overall centroid point, respectively, and $M(\eta)$ be its information matrix. Then, after some algebraic manipulation, it can be shown that

$$\varphi(\boldsymbol{\gamma}, M(\eta)) = \boldsymbol{\gamma}'\left(\sum_{i=1}^{3} D_i M^{-1}(\eta) D_i'\right)\boldsymbol{\gamma},$$

where

$$D_1 = d\begin{pmatrix} -4 & 0 & 0 & 1 & 1 & 0 \\ 0 & 2 & 0 & -2 & 0 & 1 \\ 0 & 0 & 2 & 0 & -2 & 1 \end{pmatrix}$$

$$D_2 = d\begin{pmatrix} 2 & 0 & 0 & -2 & 1 & 0 \\ 0 & -4 & 0 & 1 & 0 & 1 \\ 0 & 0 & 2 & 0 & 1 & -2 \end{pmatrix}$$

$$D_3 = d\begin{pmatrix} 2 & 0 & 0 & 1 & -2 & 0 \\ 0 & 2 & 0 & 1 & 0 & -2 \\ 0 & 0 & -4 & 0 & 1 & 1 \end{pmatrix}$$

and d is a scalar independent of the design [cf. Eqs. (7.2.7) and (7.2.8) and the discussion that follows].

As $M^{-1}(\eta)$ is positive definite, $\phi(\boldsymbol{\gamma}, M(\eta))$ will be a convex function of $\boldsymbol{\gamma}$. Hence, $\phi(\boldsymbol{\gamma}, M(\eta))$ is maximized at some boundary point of Γ. Pal and Mandal (2008) proved the following theorem:

Theorem 8.3.2 *$\phi(\boldsymbol{\gamma}, M(\eta))$ is maximized at the extreme points, viz. at* $(1, 0, 0)'$, $(0, 1, 0)'$, *and* $(0, 0, 1)'$.

By virtue of Theorem 8.3.2, it can be easily shown that $\max_{\boldsymbol{\gamma}\in\Gamma} \phi(\boldsymbol{\gamma}, M(\eta)) = \max_i d_{3i}' M^{-1}(\eta) d_{3i}$, where d_{3i}' denotes the third row vector of D_i, $i = 1, 2, 3$.

Now, $M(\eta)$ is of the form

$$M(\eta) = \begin{pmatrix} a & b & b & b & b & b \\ & a & b & b & c & b \\ & & a & c & b & b \\ & & & b & c & c \\ & & & & b & c \\ & & & & & b \end{pmatrix}$$

where $a = \frac{1}{18}(1+26\alpha_1+\frac{19}{8}\alpha_2)$, $b = \frac{1}{18}(1-\alpha_1+\frac{11}{16}\alpha_2)$, and $c = \frac{1}{81}(1-\alpha_1-\alpha_2)$.
Hence, $M^{-1}(\eta)$ is given by

$$M^{-1}(\eta) = \begin{pmatrix} e & f & f & g & g & h \\ & e & f & g & h & g \\ & & e & h & g & g \\ & & & k & m & m \\ & & & & k & m \\ & & & & & k \end{pmatrix}$$

where e, f, g, h, k, and m are as follows:

$$e = \frac{a}{(a+2b)(a+c-2b)} - \frac{2b+c}{a+2b}g, \quad f = \frac{2b}{(a+2b)(a+c-2b)} - \frac{2b+c}{a+2b}g,$$
$$h = \frac{1}{a+c-2b} + g, \qquad k = \frac{a-c}{b-c}g,$$
$$m = \frac{1}{b-c}\left[\frac{b-a}{a+c-2b} + (c-a)g\right], \quad g = \frac{ab+ac-2b^2}{(a+c_2b)(2b^2-ab-2ac+c^2)}.$$

One therefore obtains

$$\max_{\gamma\in\Gamma} \phi(\boldsymbol{\gamma}, M(\eta)) = \frac{3d^2}{\alpha_1\alpha_2(8\alpha_1+\alpha_2)}\Bigg[(16\alpha_1+\alpha_2)(8\alpha_1+3\alpha_2) + 18432\frac{27\alpha_1\alpha_2+(32\alpha_1+\alpha_2)(1-\alpha_1-\alpha_2)}{9\alpha_1\alpha_2+(16\alpha_1+\alpha_2)(1-\alpha_1-\alpha_2)}(4\alpha_1+\alpha_2)^2\Bigg],$$

which is minimized at $\alpha_1 = 0.3734$, $\alpha_2 = 0.6041$, and $\alpha_3 = 0.0225$.

Remark 8.3.2 It has been observed in Chap. 7 and Sect. 8.2 of this chapter that the optimal designs under both trace criterion and deficiency criterion assign zero mass to the overall centroid point. But in minimax method, a positive mass, however small, is assigned to the overall centroid point.

References

Chatterjee, S. K., & Mandal, N. K. (1981). Response Surface Designs for estimating the optimal point. *Bulletin of the Calcutta Statistical Association*, *30*, 145–169.

Cheng, R. C. H., Melas, V. B., & Pepelyshev, A. N. (2001). Optimal designs for the evaluation of an extremum point. In V. B. Atkinson, B. Bogacka, & A. Zhigljavsky (Eds.), *Optimum design 2000* (pp. 15–24). Dordrecht: Kluwer Academic Publishers.

Draper, N. R., & Pukelsheim, F. (1999). Kiefer ordering of simplex designs for first and second degree mixture models. *Journal of Statistical Planning and Inference*, *79*, 325–348.

Mandal, N. K., & Heiligers, B. (1992). Minimax designs for estimating the optimum point in a quadratic response surface. *Journal of Statistical Planning and Inference*, *31*, 235–244.

Mandal, N. K., & Pal, M. (2008). Optimum mixture design using deficiency criterion. *Communications in Statistics-Theory and Methods*, *37*, 565–575.

Pal, M., & Mandal, N. K. (2008). Minimax designs for optimum mixtures. *Statistics & Probability Letters*, *78*, 608–615.

Chapter 9
Optimal Designs for Estimation of Optimum Mixture in Scheffé's Quadratic Model Under Constrained Factor Space

Abstract While in Chap. 7, we have discussed determination of optimum designs for the estimation of optimum mixture when the mixing proportions vary in the whole simplex, in the present chapter we address the problem when (i) one of the proportions is bounded above, (ii) there is a cost constraint. Here, again, the trace criterion is used to find the optimum design.

Keywords Quadratic mixture models · Optimum mixing proportions · Trace optimality criterion · Pseudo-Bayesian approach · Kiefer's equivalence theorem · Optimum designs

9.1 Introduction

In many practical situations, the experimenter is faced with the problem of determining the optimum mixing proportions, when certain restrictions are placed on one or more of the components. For example, a particular ingredient may be essential to be present in the mixture in at least or at most a certain proportion. When a lower bound is specified for at least one component, the problem can be solved by introducing pseudocomponents (cf. Cornell 2002). However, when an upper bound or both lower and upper bounds are indicated for one or more components, the problem becomes too difficult to tackle. In that case, some algorithms have been proposed to find the optimum designs for estimation of the parameters, or linear functions of the parameters, of the assumed model. However, estimation of nonlinear functions of parameters poses much difficulty. In this chapter, we discuss optimum designs for estimation of the optimum mixing proportions in the cases of two and three-component mixtures under constrained experimental region. Toward this, first we will deal with one-sided constraint on only one component of the mixture. Next, we will discuss about a linear cost constraint.

B. K. Sinha et al., *Optimal Mixture Experiments*, Lecture Notes in Statistics 1028,
DOI: 10.1007/978-81-322-1786-2_9,

9.2 Optimum Mixture Designs Under Constraint on One Component

Consider the mixture model (7.2.2), where the response function $\eta(\boldsymbol{x})$ is assumed to be quadratic concave in the components $x_1, x_2, \ldots, x_q$ in the experimental region $\mathcal{X}$, given by (7.1.1).

Let, $\mathcal{X}^* \subset \mathcal{X}$ denote the contrained experimental region. The explicit form of $\mathcal{X}^*$ will depend on the type of constraint imposed. As mentioned above, we will involve only one component in the constraint. It is assumed that the point $\boldsymbol{x} = \gamma$, given by (7.2.3), at which the response function is maximized, is an interior point of $\mathcal{X}^*$.

To obtain an optimum design for estimating γ, the pseudo-Bayesian approach of Pal and Mandal (2006) can be used, with the more general assumption (7.5.1) on the prior. As in Pal and Mandal (2006), the criterion for optimal choice of design is to minimize

$$\phi(\xi) = \text{Trace } \mathcal{E}\{E[(\hat{\gamma} - \gamma)(\hat{\gamma} - \gamma)']\}$$

where $\mathcal{E}$ stands for expectation with respect to the prior of γ.

Suppose the constrained region is defined as

$$\mathcal{X}^* = \{\boldsymbol{x}\} \mid x_i \geq 0, i = 1, 2, \ldots, q; x_k \leq c_k, \text{for some } 1 \leq k \leq q, \sum_{i=1}^{q} x_i = 1\}. \quad (9.2.1)$$

Mandal et al. (2008b) discussed the optimum designs for $q = 2, 3$.

9.2.1 *Case of Two Component Mixture Model*

Suppose the constraint is imposed on the first component, that is, $x_1 \leq c$.

Since $x_1 + x_2 = 1$, invoking the result of Liski et al. (2002), and arguing as in Sect. 7.5 for finding the optimum design, one can restrict to the class $\mathcal{D}_c$ of three-point designs with support points (0, 1), $(c, 1 - c)$, and $(d_1, 1 - d_1)$, $d_1 \in (0, c)$.

For easy algebraic manipulation, one can use an alternative representation of model (7.2.2) in the two-component case:

$$\eta_x = \theta_{11} x_1 (x_1 - d_1) + \theta_{22}(x_2 - (1 - c))(x_2 - (1 - d_1)) + \theta_{12}(x_1 - c)(x_2 - 1) \quad (9.2.2)$$

where $\boldsymbol{\theta} = (\theta_{11}, \theta_{22}, \theta_{12})$ and $\boldsymbol{\beta} = (\beta_{11}, \beta_{22}, \beta_{12})$ are related by

$$\boldsymbol{\beta} = L\boldsymbol{\theta}$$

with

$$L = \begin{pmatrix} 1-d_1 & (1-c)(1-d_1) & -(1-c) \\ 0 & cd_1 & 0 \\ -d_1 & -d_1(1-c)-c(1-d_1) & c \end{pmatrix}.$$

Then, for a design $\xi \in \mathcal{D}_c$, which assigns masses α_1 to $(0, 1)$, α_2 to $(c, 1-c)$ and $1-\alpha_1-\alpha_2$ to $(d_1, 1-d_1)$, the information matrix for the vector parameter $\boldsymbol{\theta}$ is given by

$$M(\xi, \boldsymbol{\theta}) = \begin{pmatrix} \alpha_2 c^2(c-d_1)^2 & 0 & 0 \\ & \alpha_1 c^2 d_1^2 & 0 \\ & & (1-\alpha)d_1^2(c-d_1)^2 \end{pmatrix}, \qquad (9.2.3)$$

where $\alpha = \alpha_1 + \alpha_2$.

Hence, $M^{-1}(\xi, \boldsymbol{\beta}) = LM^{-1}(\xi, \boldsymbol{\theta})L'$.

Assuming unequal second-order prior moments of the components of $\boldsymbol{\gamma}$ as given in (7.5.1), we obtain the criterion function as

$$\phi(\xi) = trM^{-1}(\xi, \boldsymbol{\theta})G,$$

where $G = L'\mathcal{E}(A'(\boldsymbol{\gamma})A(\boldsymbol{\gamma}))L = ((g_{ij}))$, say, and $A(\boldsymbol{\gamma})$ is given by (7.2.8). The diagonal elements of G are given by

$$\begin{aligned} g_{11} =& 8(1-d_1)^2 v_1 + 2d_1^2(v_1+v_2-2w_{12}) - 8d_1(1-d_1)(w_{12}-v_1) \\ g_{22} =& 8(1-c)^2(1-d_1)^2 v_1 + 8c^2 d_1^2 v^2 + 2(d_1(1-c)+c(1=d_1))^2(v_1+v_2-2w_{12} \\ & - 16c(1-c)d_1(1-d_1)w_{12} - 8(1-c)(1-d_1)(d_1(1-c)+c(1-d_1))(w_{12}-v_i) \\ & - 8cd_1(d_1(1-c)+c(1-d_1))(w_{12}-v_1) \\ g_{33} =& 8(1-c)^2 v_1 + 2c^2(v_1+v_2-2w_{12}) - 8c(1-c)(w_{12}-v_1). \end{aligned}$$

Thus, for given d_1,

$$\phi(\xi) = \frac{g_{11}}{\alpha_2 c^2(c-d_1)^2} + \frac{g_{22}}{\alpha_1 c^2 d_1^2} + \frac{g_{33}}{(1-\alpha)d_1^2(c-d_1)^2} \geq \left(\sum_i \sqrt{g_{ii}^*}\right)^2, \qquad (9.2.4)$$

where

$$g_{11}^* = \frac{g_{22}}{c^2 d_1^2}, \quad g_{22}^* = \frac{g_{11}}{c^2(c-d_1)^2}, \quad g_{33}^* = \frac{g_{33}}{d_1^2(c-d_1)^2}.$$

Equality in (9.2.4) holds for

$$\alpha_i = \alpha_i^*(d_1) = \frac{\sqrt{g_{ii}^*}}{\sum_j \sqrt{g_{jj}^*}}, i = 1, 2, \tag{9.2.5}$$

The optimum value of d_1, say d_1^*, is then obtained by minimizing $\left(\sum_i \sqrt{g_{ii}^*}\right)^2$.

Thus, the optimal design assigns mass $\alpha_1^*(d_1^*)$, $\alpha_2^*(d_1^*)$ and $1-\alpha_1^*(d_1^*)-\alpha_2^*(d_1^*)$, respectively, to the support points $(0, 1)$, $(c, 1-c)$ and $(d_1^*, 1-d_1^*)$.

Remark 9.2.1 From the computations shown in Table 9.1, the following observations are made:

(a) For fixed v_1, v_2, and w_{12}, (i) min. $\phi(\xi) \downarrow c$; (ii) $d_1 \uparrow c$; (iii) $1-\alpha_1-\alpha_2$ (mass at $(d_1, 1-d_1)$) $\downarrow c$.
(b) For fixed $v_2(v_1)$ and c, min $\phi(\xi) \uparrow v_1(v_2)$.

9.2.2 Case of Three Component Mixture Model

Here, it is assumed that the prior moments satisfy

$$v_1 = v_2, w_{13} = w_{23}. \tag{9.2.6}$$

Further, suppose the third component is constrained, viz. $0 \le x_3 \le c$, where $0 < c \le 1$. To obtain the optimum design in such a situation, one can proceed as follows.

9.2.3 A Heuristic Search for Optimum Design

It is noted that under the assumption (9.2.6), the problem is invariant with respect to the first two components, and hence, the optimum design must also have the same invariance property with respect to the first two components. Now, arguing as in Sect. 7.5, the response function (7.2.2) can be represented as a quadratic in x_3, so that for the optimum design x_3 will take the three distinct values 0, c and some $a \in (0, c)$. In view of the invariant case considered in Pal and Mandal (2006) or the non-invariant case of Mandal et al. (2008a), one can restrict to the following subclass of designs with support points as given in Table 9.2, where $0 < a \le 1, a \in (0, c)$, $W_i \ge 0, i = 1, 2, 3$, $W_1 + W_2 + W_3 = 1$. Here, W_1, W_2, and W_3 denote the masses attached to $x_3 = 0, c$ and a, respectively, while the third column gives the masses for different (x_1, x_2) combinations when x_3 is given.

Table 9.1 Optimum designs and the minimum $\phi(\xi)$ for some combinations of v_1, v_2, w_{12} and c

v_1	v_2	w_{12}	c	d_1	α_2	α_1	Min. $\phi(\xi)$
0.26	0.26	0.24	1	0.5	0.3646	0.3646	17.4530
			0.8	0.3501	0.4127	0.1950	41.9121
			0.6	0.2934	0.3894	0.1256	210.158
			0.4	0.1998	0.3248	0.1765	2010.959
0.26	0.30	0.22	1	0.5092	0.3029	0.3337	29.0988
			0.8	0.3824	0.3514	0.2338	65.6210
			0.6	0.2942	0.3520	0.1710	269.3375
			0.4	0.1996	0.3148	0.1886	2189.214
0.30	0.28	0.21	1	0.4967	0.3144	0.3010	34.6172
			0.8	0.3856	0.3470	0.2151	84.9551
			0.6	0.2963	0.3403	0.1765	3555.406
			0.4	0.1998	0.3071	0.1956	2750.812
0.30	0.30	0.20	1	0.5	0.3000	0.3000	40.0000
			0.8	0.3895	0.3326	0.2250	95.9710
			0.6	0.2966	0.3311	0.1866	389.2969
			0.4	0.1997	0.3038	0.1998	2837.252
0.30	0.40	0.15	1	0.5048	0.2345	0.3008	66.0020
			0.8	0.3976	0.2694	0.2636	148.5268
			0.6	0.2982	0.2835	0.2334	510.6013
			0.4	0.1997	0.2900	0.2148	3262.091
0.30	0.48	0.11	1	0.5052	0.2435	0.3008	86.3329
			0.8	0.3997	0.2694	0.2636	189.1602
			0.6	0.2990	0.2835	0.2334	609.3615
			0.4	0.1998	0.2812	0.2240	3595.568
0.40	0.30	0.15	1	0.4950	0.2997	0.2598	66.0020
			0.8	0.3949	0.3146	0.2165	170.6315
			0.6	0.2986	0.3092	0.2016	660.2805
			0.5	0.2494	0.3006	0.2049	1570.399
0.40	0.40	0.10	1	0.5000	0.2717	0.2717	92.1051
			0.8	0.3978	0.2887	0.2378	223.0933
			0.6	0.2990	0.2918	0.2196	786.0946
			0.5	0.2496	0.2882	0.2183	1786.614
0.48	0.30	0.11	1	0.4948	0.3008	0.2435	86.3329
			0.8	0.3965	0.3086	0.2136	230.0149
			0.6	0.2991	0.3012	0.2068	881.3037
			0.5	0.2496	0.2936	0.2106	2062.470
0.48	0.48	0.02	1	0.5000	0.2650	0.2650	133.2798
			0.8	0.3989	0.2773	0.2413	323.6493
			0.6	0.2994	0.2803	0.2284	1104.629
			0.5	0.2498	0.2782	0.2270	2446.772

Table 9.2 Support points of designs in the proposed subclass

x_1	x_2	Mass	x_3	Mass
1	0	α		
0	1	α	0	W_1
1/2	1/2	$1-2\alpha$		
$(1-c)/2$	$(1-c)/2$	1	c	W_2
$1-a$	0	1/2		
0	$1-a$	1/2	a	W_3

Let a design in the above subclass be denoted by $\xi(a, \alpha, \boldsymbol{W})$. Then, after a little algebra, the information matrix for the design comes out to be

$$M(\xi) = D\Lambda D'$$

where

$$D = \begin{pmatrix} \sqrt{\alpha} & 0 & b & \frac{(1-c)^2}{4} & \frac{(1-a)^2}{\sqrt{2}} & 0 \\ 0 & \sqrt{\alpha} & b & \frac{(1-c)^2}{4} & 0 & \frac{(1-a)^2}{\sqrt{2}} \\ 0 & 0 & b & \frac{(1-c)^2}{4} & 0 & 0 \\ 0 & 0 & 0 & c^2 & \frac{a^2}{\sqrt{2}} & \frac{a^2}{\sqrt{2}} \\ 0 & 0 & 0 & \frac{c(1-c)}{2} & \frac{a(1-a)}{\sqrt{2}} & 0 \\ 0 & 0 & 0 & \frac{c(1-c)}{2} & 0 & \frac{a(1-a)}{\sqrt{2}} \end{pmatrix},$$

$\Lambda = Diag(W_1 I_3, W_2, W_3 I_2), b = \sqrt{1-2\alpha}/4$.

Therefore, the criterion function reduces to

$$\phi(\xi(a, \alpha, \boldsymbol{W})) = tr\Lambda^{-1}[D^{-1}\mathcal{E}(A'(\gamma)A(\gamma))D'^{-1}] = \frac{g_{11}^*}{W_1} + \frac{g_{22}^*}{W_2} + \frac{g_{33}^*}{W_3},$$

where

$$D^{-1}\mathcal{E}(A'(\gamma)A(\gamma))D'^{-1} = ((g_{ij})), \text{ say, } A(\gamma) \text{ is given by (7.2.8)}$$

and

$$g_{11}^* = g_{11} + g_{22} + g_{33}, g_{22}^* = g_{44}, g_{33}^* = g_{55} + g_{66},$$
$$g_{11} = g_{22}, g_{55} = g_{66}.$$

It may be noted that g_{44} and g_{55} are independent of α.

Then, for given $(\alpha, \boldsymbol{W})$, $\phi(\xi(a, \alpha, \boldsymbol{W}))$ is minimized at $\alpha = \alpha_0 = \dfrac{\sqrt{g*}}{2\sqrt{g^*} + \sqrt{2h^*}}$, where $g^* = 2ag_{11}$ and $h^* = (1 - 2a)g_{33}$.

At $\alpha = \alpha_0$,

$$\begin{aligned}\phi(\xi(a, \alpha_0, \boldsymbol{W})) \equiv \phi(\xi(a, \boldsymbol{W}) &= \frac{g^*_{11,0}}{W_1} + \frac{g^*_{22,0}}{W_2} + \frac{g^*_{33,0}}{W_3} \\ &\geq \left(\sum_i \sqrt{g^*_{ii,0}}\right)^2 \end{aligned} \tag{9.2.7}$$

where $g^*_{ii,0} = g^*_{ii} \mid_{\alpha=\alpha_0}, i = 1, 2, 3$.

Equality holds in (9.2.7) at $W_1 \equiv W_i(a) = \dfrac{\sqrt{g^*_{ii,0}}}{\sum_i \sqrt{g^*_{ii,0}}}, i = 1, 2, 3$.

Writing $W(a) = (W_1(a), W_2(a), W_3(a))'$, it therefore follows that, for given a,

$$\phi(\xi(a, \alpha, \boldsymbol{W})) \geq \phi(\xi(a, \boldsymbol{W}(a))) = (\sum_i \sqrt{g^*_{ii,0}})^2 \text{ for all } \alpha, \boldsymbol{W}. \tag{9.2.8}$$

Optimal value of a is then obtained by minimizing $\left(\sum_i \sqrt{g^*_{ii,0}}\right)^2$. However, algebraic deduction of optimal a is intractable. The optimal value a^* of a, and hence those of α and $\boldsymbol{W}$, for some combinations of $(v_1 = v_2, v_3, w_{12}, w_{13} = w_{23})$ have been computed by Mandal et al. (2008b), which are presented in Table 9.3. It may be noted that the designs obtained are optimal within the subclass $\mathcal{D}_0$ defined in (9.2.9) below:

$$\{\xi(a, \alpha, \boldsymbol{W}); 0 < a \leq 1, 0 < \alpha < 1, W_i \geq 0, i = 1, 2, 3, W_1 + W_2 + W_3 = 1\}. \tag{9.2.9}$$

To verify the optimality of $\xi(a^*, \boldsymbol{W}(a^*))$ in the entire class of competing designs, one can use Theorem 7.4.2, obtained from Kiefer's equivalence theorem. Since algebraic verification is rather difficult, Mandal et al. (2008b) checked the optimality condition (7.4.2) by taking innumerable combinations of $c, v_i, w_{ij}; i, j = 1, 2, 3$. It has been seen that for $c = 1$, the condition is satisfied for all $\boldsymbol{x}$ in $\mathcal{X}^*$. However, when c takes some value less than 1, the condition is satisfied at all points except for a very small area in $\mathcal{X}^*$. A closer look at the Table 9.4 shows that the optimality condition is violated at some of the support points of $\xi(a^*, \boldsymbol{W}(a^*))$ which indicates that more mass should be attached at those support points.

Table 9.4 shows that for given $(v_1 = v_2, v_3, w_{12}, w_{13} = w_{23})$, as c decreases, c^* deviates more from it. This indicates increase in the region of violation of the condition of optimality with decrease in c.

However, one can find the optimum design sequentially by starting with the design $\xi(a^*, \boldsymbol{W}(a^*))$ and using one of the standard algorithms, for finding optimum designs like V algorithm of Fedorov (1972).

Table 9.3 Optimal values of a, α and $\boldsymbol{W}$ for some combinations of ($v_1 = v_2, v_3, w_{12}, w_{13} = w_{23}$) and c

	Parameters				Optimal			Trace
c	$v_1 = v_2$	v_3	w_{12}	$w_{13} = w_{23}$	a	α	$\boldsymbol{W}_1$ $\boldsymbol{W}_2$ $\boldsymbol{W}_3$ 0.4455	
0.99	0.2	0.1	0.15	0.05	0.4833	0.2520	0.1095 0.4450 0.4714	491.4126
	0.2	0.1	0.12	0.065	0.4866	0.2512	0.0967 0.4319 0.4890	488.6677
	0.2	0.1	0.10	0.075	0.4905	0.2506	0.0872 0.4238 0.3793	484.5852
	0.15	0.2	0.12	0.065	0.4960	0.2508	0.1424 0.4783 0.3802	483.7883
	0.1	0.2	0.065	0.1175	0.5082	0.2210	0.1263 0.4935 0.4282	287.8529
0.9	0.2	0.1	0.15	0.05	0.4526	0.2586	0.1267 0.4451 0.4549	559.3978
	0.2	0.1	0.12	0.065	0.4579	0.2558	0.1146 0.4305 0.4890	543.4693
	0.2	0.1	0.10	0.075	0.4636	0.2541	0.1054 0.4210 0.3622	530.1226
	0.15	0.2	0.12	0.065	0.4668	0.2594	0.1671 0.4707 0.3649	567.9032
	0.1	0.2	0.065	0.1175	0.4920	0.2213	0.1617 0.4734 0.4176	316.4064
	0.2	0.1	0.15	0.05	0.4334	0.2640	0.1373 0.4451 0.4443	612.0382
	0.2	0.1	0.12	0.065	0.4394	0.2595	0.1260 0.4297 0.4634	586.1483

(continued)

Table 9.3 (continued)

	Parameters				Optimal			Trace
c	$v_1 = v_2$	v_3	w_{12}	$w_{13} = w_{23}$	a	α	$\boldsymbol{W}_1$ $\boldsymbol{W}_2$ $\boldsymbol{W}_3$ 0.4455	
0.85	0.2	0.1	0.10	0.075	0.4459	0.2568	0.1172 0.4194 0.3521	565.8259
	0.15	0.2	0.12	0.065	0.4469	0.2666	0.1812 0.4467 0.3541	637.7346
	0.1	0.2	0.065	0.1175	0.4714	0.2220	0.1854 0.4605	343.3944

Table 9.4 The upper bound c^* of x_3 so that optimality condition (7.4.2) is satisfied at all points on the plane $x_1 = 0$ or $x_2 = 0$ for the design $\xi(a^*, \boldsymbol{W}(a^*))$ for some combinations of $(v_1 = v_2, v_3, w_{12}, w_{23}, c)$

$v_1 = v_2$	w_{12}	$w_{13} = w_{23}$	v_3	c^*			
				c = 0.9	c = 0.8	c = 0.7	c = 0.5
0.2	0.15	0.05	0.1	0.8947	0.7844	0.6743	0.4602
0.15	0.2	0.12	0.065	0.8913	0.7759	0.6642	0.4565
0.1	0.2	0.065	0.1175	0.8742	0.7263	0.6104	0.4399

9.2.4 A Competitive Design

Mandal et al. (2008b) proposed another design $\xi_1(a_1, \boldsymbol{W})$, which seems to be a strong contender for the target design:

x_1	x_2	Mass	x_3	Mass
1	0	α		
0	1	α	0	W_1
1/2	1/2	1 - 2α		
1 - c	0	1/2	c	W_2
0	$1 - c$	1/2		
$1 - a_1$	0	1/2		
0	$1 - a_1$	1/2	a_1	W_3

where $0 < \alpha \leq 1, a_1 \in (0, c), W_i > 0, i = 1, 2, 3, W_1 + W_2 + W_3 = 1$ and W_is and $\boldsymbol{\alpha}$ are defined as before. For the design $\xi_1(a_1, \alpha, \boldsymbol{W})$, suppose one uses the same masses α, W_1, W_2, and W_3 as in the optimum design for the unconstrained

Table 9.5 The design $\xi_1(a_1^*, \alpha, W)$ with the optimum design $\xi(a^*, W(a^*))$ in the subclass D_0 given by (9.2.9)

Parameters				Values of		c	$\xi_1(a_1^*, \alpha, \mathbf{W})$		Trace of
$v_1 = v_2$	v_3	w_{12}	$w_{13} = w_{23}$	α	W_1 W_2 W_3		a_1^*	Trace	$\xi(a^*, \mathbf{W}^*(a^*))$
0.15	0.2	0.12	0.065	0.2501	0.4897	0.99	0.4954	483.7396	483.7883
						0.9	0.4577	564.5837	567.9032
					0.0863	0.85	0.4319	631.9667	637.7346
						0.7	0.3465	1054.5752	1066.964
					0.4240	0.5	0.2379	3639.1909	3836.257
0.2	0.1	0.15	0.05	0.2514	0.4473	0.99	0.4827	491.3899	491.4126
						0.9	0.4448	558.0290	559.3978
					0.1077	0.85	0.4211	610.2784	612.0382
						0.7	0.3424	904.2737	955.737
					0.4450	0.5	0.2346	2464.5404	2429.956

experimental region involving three-component mixture, considered in Sect. 7.5, with the mass W_2 divided equally among the two points $(1 - c, 0, c)$ and $(0, 1 - c, c)$. The optimum a_1, denoted by a_1^*, is determined by minimizing the criterion function. The relative performance of the two designs is given in Table 9.5.

Remark 9.2.2 It is observed that, in terms of the criterion function, the two designs are very close to one other. So, starting with any one of these designs, one may use a standard numerical algorithm to reach the optimum design.

9.3 Optimum Designs Under Cost Constraint

In industrial experiments, cost plays an important role in deciding upon the experimental design to be chosen. Fedorov and Hackl (1997) have discussed the problem of determining an optimum design when the cost of performing the experiment is given as a function of the design point, subject to a bound on the total cost. Pal and Mandal (2009) obtained designs for estimation of optimum factor combination in the context of response surface when the costs per unit of the factors are given, subject to a bound on the total cost for a factor combination. They also obtained optimum proportions under similar constraint for the quadratic mixture model.

Consider the quadratic mixture model (7.2.2). Let $c_i, i = 1, 2, \ldots, q$, denote the cost per unit for component i, and c_0 be the total assigned cost per unit of the mixture. Then, the cost constraint on a mixture combination is given by

$$c_1x_1 + c_2x_2 + \cdots + c_kx_k = c_0 \tag{9.3.1}$$

It should be noted that for an assigned total cost, the general form of the constraint is $c_1x_1+c_2x_2+\cdots+c_kx_k \leq c_0$. However, from experimental point of view, utilization of the total assigned cost is likely to give more information. In other words, we must attain 'equality' as in (9.3.1).

Under the assumption of concavity of the response function with respect to $\boldsymbol{x}$ and that its point of maxima occurs at an interior point of the experimental region, the unconstrained maximum of the mean response $\eta(\boldsymbol{x})$ occurs at $\boldsymbol{x} = \gamma$, where γ is given by (7.2.4). If γ satisfies the constraint (9.3.1), it becomes the optimum mixing proportion for the constrained maximization problem. Else, one may proceed as follows.

Since the mixing proportions $x_1, x_2, \ldots, x_q$ must satisfy two constraints, viz. (9.3.1) and

$$x_1 + x_2 + \cdots + x_q = 1, \tag{9.3.2}$$

one can express the mean response function in terms of $(q - 2)$ independent components, with their proportions lying in the interval [0, 1].

9.3.1 Case of Two Components

Here, the proportions x_1 and x_2 are determined uniquely by the two linear constraints (9.3.1) and (9.3.2). Hence, the question of choice of x_1 and x_2 for maximizing the mean response function does not arise.

9.3.2 Case of Three Components

Using (9.3.1) and (9.3.2), one can write

$$x_2 \equiv x_2(x_1) = \frac{c_0 - c_3}{c_2 - c_3} - \frac{c_1 - c_3}{c_2 = c_3}x_1, x_3 \equiv x_3(x_1) = \frac{c_0 - c_2}{c_3 - c_2} - \frac{c_1 - c_2}{c_3 - c_2}x_1. \tag{9.3.3}$$

In order that $x_2, x_3 \in [0, 1]$, it is essential that x_1 lies in some interval $[a, b]$, where $0 \leq a < b \leq 1$ are functions of c_0, c_i, $i = 1, 2, 3$.

Hence, the response function (7.2.2) becomes

$$\eta(x) = \kappa_0 + \kappa_1 x_1 + \kappa_2 x_1^2, \tag{9.3.4}$$

where

$$\kappa_0 = \lambda_0' B \lambda_0, \quad \kappa_1 = 2\lambda_0' B \lambda_1, \quad \kappa_2 = \lambda_1' B \lambda_1,$$
$$\lambda_0 = \left(0, \frac{c_0 - c_3}{c_2 - c_3}, \frac{c_0 - c_2}{c_3 - c_2}\right), \quad \lambda_1 = \left(1, \frac{c_1 - c_3}{c_2 - c_3}, -\frac{c_1 - c_2}{c_3 - c_2}\right).$$

Clearly, $\eta(\boldsymbol{x})$ is a quadratic concave function of x_1 since B is negative definite. Hence, it is maximized at $x_1 = \gamma_1 = -\dfrac{\kappa_1}{2\kappa_2}$. As B is unknown, one can estimate B from a given design and hence obtain an estimate $\hat{\gamma}_1$, say, of γ_1.

A rational criterion for finding the optimal design would be to minimize var($\hat{\gamma}_1$). The standard delta method gives

$$\text{var}\,(\hat{\gamma}_1) = \frac{1}{4\kappa_2^2}(0, 1, -2\gamma_1)M^{-1}(\xi, \boldsymbol{\kappa})(0, 1, -2\gamma_1)',$$

where $M(\xi, \boldsymbol{\kappa})$ is the information matrix of a design ξ for estimating $\boldsymbol{\kappa} = (\kappa_0, \kappa_1, \kappa_1)'$ and is given by

$$M(\xi, \boldsymbol{\kappa}) = \begin{pmatrix} 1 & \mu_1 & \mu_2 \\ & \mu_2 & \mu_3 \\ & & \mu_4 \end{pmatrix},$$

$\mu_i = E_\xi(x_1^i), i = 1(1)4$.

Since γ_1 is a nonlinear function of the unknown parameters, var($\hat{\gamma}_1$) will depend on them. To resolve this problem, one can adopt the pseudo-Bayesian approach of Pal and Mandal (2006). It may be noted that one can disregard $1/4\kappa_2^2$ in var($\hat{\gamma}$) in the search for the optimum design.

Let the first two prior moments of γ_1 be given by $E(\gamma_1) = u$, $E(\gamma_1^2) = v$, where $u^2 < v < u$, and $u \in [0, 1]$. So, the trace optimality criterion will be

$$\phi(\xi) = \text{ Trace } M^{-1}(\xi, \boldsymbol{\kappa})\mathcal{E}[(0, 1, -2\gamma_1)'(0, 1, -2\gamma_1)], \tag{9.3.5}$$

where $\mathcal{E}$ denotes the expectation with respect to the prior distribution of γ_1. From Liski et al. (2002), it follows that the optimal design maximizing the moment matrix in the sense of Loewner order dominance puts mass at three distinct values of x_1, viz. a, b and $s \in (a, b)$. Thus, to find the design minimizing (9.3.5), one can concentrate on the class of three-point designs given by

$$\Delta(\xi) = \{(a, s, b) : (\alpha_1, 1 - \alpha_1 - \alpha_2, \alpha_2)\},$$

where α_1, $1-\alpha_1-\alpha_2$ and α_2 are the masses attached to $x_1 = a$, s and b, respectively.

Now, for easy algebraic manipulation, model (9.2.2) can be expressed as

$$\eta(x_1) = \theta_{11}(x_1 - a)(x_1 - s) + \theta_{22}(x_1 - s)(x_1 - b) + \theta_{33}(x_1 - a)(x_1 - b), \tag{9.3.6}$$

where $\boldsymbol{\theta} = (\theta_{11}, \theta_{22}, \theta_{33})$ and $\boldsymbol{\kappa} = (\kappa_0, \kappa_1, \kappa_2)$ are related by

$$\boldsymbol{\kappa} = L\boldsymbol{\theta},$$

with

$$L = \begin{pmatrix} as & bs & ab \\ -(a+s) & -(b+s) & -(a+b) \\ 1 & 1 & 1 \end{pmatrix}.$$

Then, for any design ξ, the information matrices $M(\xi, \boldsymbol{\kappa})$ and $M(\xi, \boldsymbol{\theta})$ for $\boldsymbol{\kappa}$ and $\boldsymbol{\theta}$, respectively, satisfy

$$M^{-1}(\xi, \boldsymbol{\kappa}) = LM^{-1}(\xi, \boldsymbol{\theta})L',$$

where

$$M^{-1}(\xi, \boldsymbol{\theta}) = \text{Diag}\left(\frac{1}{\alpha_2(b-a)^2(b-s)^2}, \frac{1}{\alpha_1(a-b)^2(a-s)^2}, \frac{1}{(1-\alpha_1-\alpha_2)(s-a)^2(s-b)^2}\right).$$

Thus, one gets,

$$\begin{aligned} \phi(\xi) &= \text{Trace } M^{-1}(\xi, \boldsymbol{\theta})G \\ &= \frac{1}{4\kappa_2^2}\left(\frac{g_{11}}{\alpha_2(b-a)^2(b-s)^2} + \frac{g_{22}}{\alpha_1(a-b)^2(a-s)^2} \right. \\ &\quad \left. + \frac{g_{33}}{(1-\alpha_1-\alpha_2)(s-a)^2(s-b)^2}\right), \end{aligned} \tag{9.3.7}$$

where $G = ((g_{ij}))$, with diagonal elements

$$\begin{aligned} g_{11} &= 4v + 4u(a+s) + (a+s)^2 \\ g_{22} &= 4v + 4u(b+s) + (b+s)^2 \\ g_{33} &= 4v + 4u(a+b) + (a+b)^2. \end{aligned}$$

Lower bound to (9.3.7) is given by $(\Sigma\sqrt{g_{ii}^*})^2$, where

$$g_{11}^* = \frac{g_{11}}{(b-a)^2(b-s)^2}, \quad g_{22}^* \frac{g_{22}}{(a-b)^2(a-s)^2}, \quad g_{33}^* \frac{g_{33}}{(s-a)^2(s-b)^2},$$

Table 9.6 Optimal values of s, α_1, α_2 and the minimum value ϕ^* of the criterion function for some combinations of (u, v)

u	v	s	α_1	α_2	$1-\alpha_1-\alpha_2$	ϕ^*
0.4	0.20	0.1667	0.2569	0.1976	0.5455	9084.66
0.4	0.30	0.1667	0.2339	0.1907	0.5754	13318.21
0.6	0.40	0.1667	0.2139	0.1730	0.6131	21747.41
0.6	0.50	0.1667	0.2018	0.1679	0.6303	27649.91
0.8	0.65	0.1667	0.1843	0.1554	0.6603	42433.85
0.8	0.70	0.1667	0.1822	0.1518	0.6660	46240.68

and the bound is attained for $\alpha_1 = \alpha_1^*(s) = \dfrac{\sqrt{g_{22}^*}}{\sum_i \sqrt{g_{ii}^*}}$, $\alpha_2 = \alpha_2^*(s) = \dfrac{\sqrt{g_{11}^*}}{\sum_i \sqrt{g_{ii}^*}}$.

If s^* be the value of s minimizing $(\Sigma\sqrt{g_{ii}^*})^2$, then the optimal design assigns masses $\alpha_1^*(s^*)$, $\alpha_2^*(s^*)$ and $1-\alpha_1^*(s^*)-\alpha_2^*(s^*)$, respectively, at $x_1 = a, b$, and s^*.

Thus, the optimum mixture design that maximizes the expected response, subject to the cost constraint (9.3.1), has support points at $(a, x_2(a), x_3(a))$, $(b, x_2(b), x_3(b))$ and $(s^*, x_2(s^*), x_3(s^*))$ with masses $\alpha_1^*(s^*)$, $\alpha_2^*(s^*)$ and $1-\alpha_1^*(s^*)-\alpha_2^*(s^*)$, respectively, where expressions of $x_2(\cdot)$ and $x_3(\cdot)$ are given by (9.3.3). Table 9.6 gives the optimum designs for some combinations of the prior moments (u, v) in a three-component mixture experiment, when the costs per unit of the components are $c_1 = 1, c_2 = 3, c_3 = 7$, and the total cost per unit of the mixture is restricted at $c_0 = 5$. From (9.3.3), we get $a = 0, b = 1/3$.

Remark 9.3.1 The table shows that ϕ^* increases with increase in the value of v. This is in agreement with the fact that more accurate the information on γ_1 is, smaller is the covariance matrix in the sense of trace criterion.

References

Cornell, J. A. (2002). *Experiments with mixtures* (3rd ed.). New York: Wiley.

Fedorov, V. V. (1972). *Theory of optimal experiments*. New York: Academic.

Fedorov, V. V., & Hackl, P. (1997). Model-oriented design of experiments (Vol. 125). Lecture notes in statistics. New York: Springer.

Liski, E. P., Mandal, N. K., Shah, K. R., & Sinha, B. K. (2002). *Topics in optimal design* (Vol. 163). Lecture notes in statistics New York: Springer.

Mandal, N. K., Pal, M., Sinha, B. K., & Das, P. (2008a). Optimum mixture designs: A Pseudo-Bayesian approach. *Journal of the Indian Society of Agricultural Statistics*, *62*, 174–182. (special volume in memory of Prof. K. Kishen).

Mandal, N. K., Pal, M., Sinha, B. K., & Das, P. (2008b). Optimum mixture designs under constraints on mixing components. *Statistics and Applications*, *6*, 189–205.

Pal, M., & Mandal, N. K. (2006). Optimum designs for optimum mixtures. *Statistics & Probability Letters*, *76*, 1369–1379.

Pal, M., & Mandal, N. K. (2009). Optimum designs for estimation of optimum point under cost constraint. *Journal of Applied Statistics*, *36*, 893–906.

Chapter 10
Optimal Designs for Estimation of Optimum Mixture Under Darroch–Waller and Log-Contrast Models

Abstract This chapter addresses the problem of finding optimum designs for the estimation of optimum mixture combination when the mean response is defined by (i) the additive quadratic mixture model due to Darroch and Waller (1985) and (ii) the quadratic log-contrast model due to Aichison and Bacon-Shoane (1984). Both the models have some advantage over Scheffé quadratic mixture model, in specific situations.

Keywords Quadratic mixture models · Darroch–Waller model · Mixture-amount model · Log-contrast model · Optimum mixing proportions · A-optimality criterion · D-optimality criterion · Pseudo-Bayesian approach · Kiefer's equivalence theorem · Optimum designs

10.1 Introduction

Several types of mixture models have been developed to describe the relationship between the mean response $\eta_{\boldsymbol{x}}$ and the mixing proportions $\boldsymbol{x} = (x_1, x_2, \ldots, x_q)$ where $\boldsymbol{x}$ lies in the $(q-1)$-dimensional simplex $\mathcal{X}$ which is now denoted by S^{q-1}:

$$S^{q-1} = \left\{(x_1, x_2, \ldots, x_q) : x_i \geq 0, i = 1, 2, \ldots, q, \sum_{i=1}^{q} x_i = 1\right\}, q \geq 2. \tag{10.1.1}$$

The commonly used model is the full quadratic mixture model of Scheffé (1958):

$$\eta(\boldsymbol{x}) = \sum_{i=1}^{q} \beta_i x_i + \sum_{i<j=1}^{q} \beta_{ij} x_i x_j \tag{10.1.2}$$

B. K. Sinha et al., *Optimal Mixture Experiments*, Lecture Notes in Statistics 1028, DOI: 10.1007/978-81-322-1786-2_10, © Springer India 2014

An additive quadratic model for $\eta(\boldsymbol{x})$ was studied by Darroch and Waller (1985) for the case of $q = 3$, which, in the case of q-component mixture, has the form

$$\eta(\boldsymbol{x}) = \sum_{i=1}^{q} \alpha_i\, x_i + \sum_{i=1}^{q} \alpha_{ii}\, x_i\, (1 - x_i). \tag{10.1.3}$$

For $q = 3$, the models (10.1.2) and (10.1.3) are equivalent, but for $q = 2$, the parameters of the latter model are not uniquely determined. For $q \geq 4$, (10.1.3) is a special case of (10.1.2), with the coefficients of (10.1.2) being governed by a system of linear constraints. The model (10.1.3) is additive in $x_1, x_2, \ldots, x_q$ and has fewer parameters than (10.1.2) when $q \geq 4$. It is also often found to fit data well (see Chan 2000).

Aitchison and Bacon-Shone (1984) proposed the quadratic log-contrast model given by

$$\eta(\boldsymbol{x}) = \beta_0 + \sum_{i=1}^{q-1} \beta_i \,\log(x_i/x_q) + \sum_{i=1}^{q-1} \sum_{i<j}^{q-1} \beta_{ij} \,\log(x_i/x_q)\log(x_j/x_q), \tag{10.1.4}$$

The advantage of a log-contrast model lies in the fact that as $z_i = \log(x_i/x_q)$ can be varied independently, the polynomial forms in z_is can be full in the sense of including all terms of appropriate degree, as against Scheffé (1958) polynomial models in x_is, which require the omission of certain terms to ensure identifiability.

Optimum designs for parameter estimation in models (10.1.3) and (10.1.4) have been reviewed in Chap. 6. In this chapter, we focus on optimum designs for estimation of the optimum mixture combination in Darroch–Waller and log-contrast models.

10.2 Optimality Under Darroch–Waller Model

Rewriting the model as $\eta(\boldsymbol{x}) = \boldsymbol{x}'B\boldsymbol{x}$ with the help of the constraint $\sum_{i=1}^{q} x_i = 1$, and, assuming that B is negative definite and the optimum point $\gamma = \delta^{-1}B^{-1}\mathbf{1}_q$, maximizing the response, occurs in the interior of (10.1.1), where $\delta = \mathbf{1}'B^{-1}\mathbf{1}$, Pal et al. (2012) attempted to find an A-optimal design for estimating γ. For any continuous design ξ, the dispersion matrix of the estimate $\hat{\gamma}$ is given by $D(\hat{\gamma}) \cong A(\gamma)M^{-1}(\xi)A'(\gamma)$, where M is the information matrix of ξ, and $A(\gamma)$ is the matrix of partial derivatives of γ with respect to the parameters of (10.1.3). Clearly, $D(\hat{\gamma})$ is dependent on the model parameters, and they adopted the pseudo-Bayesian approach of Pal and Mandal (2006) to do away with the nuisance parameters. Arguing as in Pal and Mandal (2006), it is assumed that $\mathcal{E}(\gamma_i^2) = v, i = 1, 2, \ldots, q; \mathcal{E}(\gamma_i\gamma_j) = w, i \neq j = 1, 2, \ldots, q; v > 0, w > 0$, where $v + (q-1)w = \frac{1}{q}, \frac{1}{q^2} < v < \frac{1}{q}, v > w > 0$. The criterion function for the optimal choice of design is, therefore,

$\phi(\xi) = \mathcal{E}[D(\hat{\gamma})] = TrM^{-1}(\xi)\mathcal{E}(A'(\gamma)A(\gamma))$. Mimicking the argument of Atwood (1969), Pal et al. (2012) proved the following theorem:

Theorem 10.2.1 *The barycenters of the experimental region* (10.1.1) are the possible support points of the A-optimal design.

Thus, one needs to restrict to the class of weighted centroid designs (WCDs) to find the optimal design. Recall that WCDs comprise of all the barycenters of the simplex.

For a q-component model, let ξ be a WCD with weight w_r at each of the barycenters of depth $(r-1)$, $1 \leq r \leq q$. Let $\xi_{1,i}^{(q)}$ be a WCD which assigns weight $w_r = 0$ for all r, except $r = 1, i, 2 \leq i \leq q$, and by $\xi_{1,i,k}^{(q)}$, we denote a WCD with weight $w_r = 0$ for all r, except $r = 1, i, k, 2 \leq i < k \leq q$. Starting with the subclass of WCDs $\xi_{1,i}^{(q)}$, it is easy to check that the optimum design within the subclass has weights

$$\left.\begin{aligned} w_1 = w_{10} &= \frac{\sqrt{d_{1i}}}{\sqrt{C(q,1)}\sqrt{C(q,1)d_{1i}} + \sqrt{C(q,i)d_{2i}}} \\ w_i = w_{i0} &= \frac{\sqrt{d_{2i}}}{\sqrt{C(q,i)}\sqrt{C(q,1)d_{1i}} + \sqrt{C(q,i)d_{2i}}} \end{aligned}\right\} \tag{10.2.1}$$

where

$$d_{1i} = q^2(q-1) + \frac{4i}{i-1}q^2(q-1)\left(\frac{1}{q} - \frac{1}{2}\right) + 4\left(\frac{i}{i-1}\right)^2 aq$$

$$d_{2i} = \left[\frac{4i^4 a}{(i-1)^2}q - \frac{4i^3}{(i-1)(q-1)}q\{a - b + bq\}\right] / C(q-2, i-1)$$

$$a = q(q-1)[v + (1/4 - 1/q)], b = q[w + (1/4 - 1/q)].$$

Pal et al. (2012) established the following results with the help of the equivalence theorem:

Theorem 10.2.2 *For* $q = 3$, *the design* $\xi_{1,2}^{(3)}$, *with* $w_1 = w_{10}$ *and* $w_2 = w_{20}$, *given by* (10.2.1), is A-optimal, whatever be $v \in \left(\frac{1}{9}, \frac{1}{3}\right)$.

Theorem 10.2.3 *For* $q = 4$, *the design* $\xi_{1,2}^{(4)}$, *with* $w_1 = w_{10}^{(2)}$ *and* $w_2 = w_{20}^{(2)}$, *given by* (10.2.1), is A-optimal when $v < v_0$, where v_0, rounded off to seven places of decimal, is 0.1975663.

For $v \geq v_0$, Pal et al. (2012) showed through numerical computation that the optimum design belongs to the subclass of WCDs designs $\xi_{1,2,3}^{(4)}$. Also, for $q = 5$, the A-optimal design has been numerically demonstrated by the authors as belonging to the subclass of WCDs designs $\xi_{1,2,3}^{(5)}$.

The A-optimal designs obtained by Pal et al. (2012) for $q = 3, 4, 5$ for some values of v are presented in Table 10.1. Note that here we are referring to γ estimation in an optimal manner and v refers to a prior parameter viz, $v = \mathcal{E}(\gamma_i^2)$ for $i = 1, 2, \ldots, q$.

Table 10.1 A-optimal designs for some values of v when $q = 3, 4, 5$

q	v	$C(q,1)w_1$	$C(q,2)w_2$	$C(q,3)w_3$	$C(q,4)w_4$	$C(q,5)w_5$
3	0.12	0.2563	0.7437	0	–	–
	0.16	0.3285	0.6715	0	–	–
	0.20	0.3490	0.6510	0	–	–
	0.24	0.3589	0.6411	0	–	–
	0.28	0.3648	0.6352	0	–	–
	0.32	0.3686	0.6314	0	–	–
4	0.08	0.2168	0.7832	0	0	–
	0.10	0.2697	0.7303	0	0	–
	0.15	0.3210	0.6790	0	0	–
	0.19	0.3389	0.6611	0	0	–
	0.20	0.3419	0.6566	0.0015	0	–
	0.22	0.3454	0.6425	0.0120	0	–
	0.24	0.3484	0.6308	0.0208	0	–
5	0.06	0.1999	0.6403	0.1598	0	0
	0.08	0.2333	0.5113	0.2554	0	0
	0.10	0.2502	0.4184	0.3314	0	0
	0.12	0.2602	0.3461	0.3937	0	0
	0.14	0.2666	0.2876	0.4458	0	0
	0.16	0.2710	0.2392	0.4898	0	0
	0.18	0.2741	0.1977	0.5282	0	0

Remark 10.2.1 It may be mentioned that for parameter estimation in the Darroch–Waller model, the optimum design for $q = 3, 4$ has support points only at barycenters of depths 0 and 1, while for $q = 5$, the support points are at the barycenters of depths 0 and 2 (cf. Chan et al. 1998a).

A modification of Darroch–Waller model to include the amount of mixture has been suggested by Zhang et al. (2005). Let A denote the maximum possible amount of the mixture. If $a_i (\geq 0), i = 1, 2, \ldots, q$, denote the actual amounts of the q components in the mixture, then $\sum_{i=1}^{q} a_i \leq A$. The proportion of the ith component in the mixture is $x_i = a_i/A$, $(i = 1, 2, \ldots, q)$, and the component space is given by

$$\Xi = \{\boldsymbol{x} = (x_1, x_2, \ldots, x_q) : x_i \geq 0, i = 1, 2, \ldots, q,\ x_1 + x_2 + \cdots + x_q \leq 1\}. \tag{10.2.2}$$

The response function $\eta(\boldsymbol{x})$ is defined as

$$\eta(\boldsymbol{x}) = \alpha_0 + \sum_{i=1}^{q} \alpha_i x_i + \sum_{i=1}^{q} \alpha_{ii} x_i (1 - x_i). \tag{10.2.3}$$

It has been pointed out by Zhang et al. (2005) that (10.2.3) remains the same even when the maximum amount A is replaced by another maximum amount A^*.

The model is a quadratic model, additive in the mixing proportions $x_1, x_2, \ldots, x_q$, and, for $q \geq 4$, it has fewer parameters than the mixture–amount model obtained from Scheffé (1958) full quadratic model, modified by inclusion of a constant term β_0, viz.

$$\eta(\boldsymbol{x}) = \beta_0 + \sum_{i=1}^{q} \beta_i x_i + \sum_{i<j=1}^{q} \beta_{ij} x_i x_j.$$

Mandal et al. (2012) investigated A-optimal designs for estimating the optimum mixture combination. Let $x_{q+1} = 1 - \sum_{i=1}^{q} x_i$. Then, x_{q+1} represents the proportion of the total amount that is not used in the mixture and satisfies $x_{q+1} \geq 0, \sum_{i=1}^{q+1} x_i = 1$. Defining $\boldsymbol{x}^* = (x_1, x_2, \ldots, x_{q+1})'$ and using the constraint $\sum_{i=1}^{q+1} x_i = 1$, the model (10.2.3), can be, as before, written as $\eta(\boldsymbol{x}^* = \boldsymbol{x}^{*\prime} B \boldsymbol{x}^*)$, where B is a symmetric matrix, involving the model parameters, and is assumed to be negative definite. The optimum point, maximizing the response, is given by $\gamma^* = (\gamma_1, \gamma_2, \ldots, \gamma_q, \gamma_{q+1})' = \delta^{*-1} B^{-1} \mathbf{1}_{q+1}$, where $\delta^* = \mathbf{1}_{q+1} B^{-1} \mathbf{1}_{q+1}$, under the assumption that the optimum point occurs in the interior of the experimental region $S^q = \{(x_1, x_2, \ldots, x_q, x_{q+1})' : x_i \geq 0, i = 1, 2, \ldots, q+1, \sum_{i=1}^{q+1} x_i = 1\}$.

Writing $S_\delta^q = \{\boldsymbol{x} = (x_1, x_2, \ldots, x_q) : x_i \geq 0, i = 1, 2, \ldots, q, \sum_{i=1}^{q} x_i = \delta\}$, $\delta \in [0, 1]$, the component space (10.2.2) is given by $\Xi = \bigcup_{\delta \in [0,1]} S_\delta^q$.

Using the argument of Zhang et al. (2005), Mandal et al. (2012) showed that only the barycenters of S_δ^q, $\delta \in [0, 1]$, play a crucial role in the construction of the optimum designs. For $6 \leq q \leq 10$, the following theorem has been established:

Theorem 10.2.4 *For* $6 \leq q \leq 10$, *the A-optimal design assigns positive weights only at the barycenters of depths* 0, 1, and 2 in S_1^q, whatever be the prior moments.

To prove the theorem, Mandal et al. (2012) made use of Kiefer's equivalence theorem. For $3 \leq q \leq 5$, they verified, using equivalence theorem, that the optimal design does not necessarily belong to the subclass of designs with support points only at the barycenters of S_1^q. Since algebraic derivation is rather involved, they then proceeded numerically to find the optimum designs for some combinations of the prior moments. The following table showing A-optimal designs is reproduced from Pal et al. (2012), where w_0 is the weight assigned to the origin $\mathbf{0} = (0, 0, \ldots, 0)'$, w_{j+1} the weight assigned to each of the barycenters of depth j in S_1^q, and w_δ the weight assigned to each barycenter of depth 0 in S_δ^q, for some $\delta \in (0, 1)$, and e_1, v_1 are the common first- and second-order prior moments of γ_i, $i = 1, 2, \ldots, q$ (Table 10.2).

10.3 Optimality Under Log-Contrast Model

Chan (1992) discussed the D-optimal design for parameter estimation in (10.1.4) with experimental domain restricted to (10.3.1) given below. Huang and Huang (2009) also studied D_s-optimal designs for discriminating between linear and

Table 10.2 A-optimal designs for some combinations of the prior moments of γ

q	e_1	v_1	δ	w_0	$C(q,1)w_1$	$C(q,2)w_2$	$C(q,3)w_3$	$C(q,1)w_\delta$
3	0.10	0.05	0.3742	–	0.200378	0.509256	–	0.290366
	0.25	0.16	0.3741	–	0.289217	0.548943	–	0.161842
	0.30	0.22	0.3696	–	0.326449	0.539221	–	0.134332
4	0.10	0.05	0.3458	–	0.235803	0.680893	–	0.083307
	0.15	0.10	0.3446	–	0.272932	0.651472	–	0.075595
	0.20	0.10	0.3458	–	0.267992	0.647196	–	0.084813
5	0.10	0.05	0.3461	–	0.186775	0.244744	0.51214	0.056341
	0.15	0.06	0.3590	–	0.200957	0.378069	0.35637	0.064604
	0.18	0.10	0.3568	–	0.237630	0.324340	0.37891	0.059122
6	0.06	0.01	–	0.018950	0.100794	–	0.880255	–
	0.10	0.05	–	0.017466	0.173260	–	0.809274	–
	0.14	0.06	–	0.018844	0.179778	–	0.801378	–
7	0.06	0.01	–	0.015482	0.121510	–	0.863009	–
	0.10	0.02	–	0.016845	0.119170	–	0.863986	–
	0.12	0.05	–	0.015451	0.176783	–	0.807766	–
8	0.06	0.01	–	0.013165	0.125501	–	0.861335	–
	0.08	0.02	–	0.013330	0.138480	–	0.848190	–
	0.10	0.05	–	0.012364	0.186361	–	0.801275	–
9	0.06	0.01	–	0.011343	0.128415	–	0.860242	–
	0.08	0.02	–	0.011480	0.141626	–	0.846894	–
	0.10	0.05	–	0.010626	0.190205	–	0.799169	–
10	0.03	0.01	–	0.009038	0.149193	–	0.841769	–
	0.06	0.02	–	0.009401	0.155192	–	0.835406	–
	0.09	0.05	–	0.009006	0.194969	–	0.796025	–

quadratic log-contrast models by restricting to the experimental region $\mathcal{X}_\delta = \{(x_1, x_2, \ldots, x_q)' \in \text{rel. int. } S^{q-1} : \delta \leq x_i/x_j \leq 1/\delta, \text{ for all } i = 1, 2, \ldots, q\}, \delta \in (0, 1)$. Following in their footsteps, Pal and Mandal (Pal and Mandal, 2012c), in order to estimate the optimum mixture combination, restricted the experimental region to

$$S_\delta^{q-1} = \{(x_1, x_2, \ldots, x_q)' \in \text{rel. int. } S^{q-1} : \delta \leq x_i/x_q \leq 1/\delta, i = 1, 2, \ldots, q-1\}, \delta \in (0, 1), \tag{10.3.1}$$

where rel. int. S^{q-1} denotes the relative interior of S^{q-1}.

Setting $t_i = -\log\left(\frac{x_i}{x_q}\right)/\log\delta$, model (10.1.4) can be written as

$$\eta(\boldsymbol{x}) \equiv \eta(\boldsymbol{t}) = \lambda_0 + \sum_{i=1}^{q-1} \lambda_i t_i + \sum_{i=1}^{q-1}\sum_{j=1}^{q-1} \lambda_{ij} t_i t_j = \boldsymbol{f}'(\boldsymbol{t})\boldsymbol{\lambda}^*, \text{ say} \tag{10.3.2}$$

where

$$\begin{aligned}
\boldsymbol{f}(\boldsymbol{t}) &= (1, t_1, t_2, \ldots, t_{q-1}, t_1^2, t_2^2, \ldots, t_{q-1}^2, t_1t_2, t_1t_3, \ldots, t_{q-2}t_{q-1})',\\
\boldsymbol{\lambda}^* &= (\lambda_0, \lambda_1, \ldots, \lambda_{q-1}, \lambda_{11}, \lambda_{22}, \ldots, \lambda_{q-1,q-1}\lambda_{12}, \lambda_{13}, \ldots, \lambda_{q-2,q-1})',\\
\lambda_0 &= \beta_0, \lambda_i = \beta_i(-\log\delta), \lambda_{ij} = \beta_{ij}(\log\delta)^2,
\end{aligned}$$

and the experimental domain in terms of $\boldsymbol{t}$ becomes

$$\begin{aligned}
\mathcal{F} = \{&\boldsymbol{t} = (t_1, t_2, \ldots, t_{q-1})' \in [-1, 1]^{q-1} :\\
&t_i - t_j \in [-1, 1] \text{ for all } i, j = 1, 2, \ldots, q-1\}.
\end{aligned} \tag{10.3.3}$$

The observations y_is on the response are assumed to be independent, with constant variance σ^2. Though the choice of the divisor x_q in t_i is arbitrary, for the sake of convenience, one can take x_q, the last component, as the divisor.

To find the optimum design for estimating the optimum $\boldsymbol{x}$ in the domain (10.3.1) that maximizes $\eta_{\boldsymbol{x}}$, one can, therefore, first find the optimum design for estimating optimum $\boldsymbol{t}$ in the domain (10.3.3), which maximizes $\eta(\boldsymbol{t})$. To do so, it is convenient to write the model (10.3.2) as

$$\eta(\boldsymbol{t}) = \lambda_0 + \boldsymbol{\lambda}'\boldsymbol{t} + \boldsymbol{t}'\Delta\boldsymbol{t}, \tag{10.3.4}$$

where $\boldsymbol{\lambda} = (\lambda_1, \lambda_2, \ldots, \lambda_{q-1})'$, $\boldsymbol{t} = (t_1, t_2, \ldots, t_{q-1})'$, Δ is a $(q-1) \times (q-1)$ matrix of the form $\Delta = \frac{1}{2}((1+\delta_{ij})\lambda_{ij})$, $\lambda_{ji} = \lambda_{ij}$, for $j > i$, and δ_{ij} is the Kronecker delta taking the value 1 if $i = j$, and 0 if $i \neq j$. It is assumed that Δ is negative definite and the optimum t is an interior point of the experimental domain $\mathcal{F}$. The optimum point in (10.3.3) is then obtained as $\gamma = -\frac{1}{2}\Delta^{-1}\boldsymbol{\lambda}$, which is a nonlinear function of the parameters of (10.3.2). From the inverse transformation $\boldsymbol{t} \to \boldsymbol{x}$, it is easy to find $x = x_0$, which maximizes (10.1.4), and this is also a nonlinear function of the model parameters.

For a (continuous) design ξ, the large sample dispersion of the estimate $\hat{\gamma}$ of γ will be, as before, given by $D(\hat{\gamma}) \cong A(\gamma)M^{-1}(\xi)A'(\gamma)$, where M is the information (moment) matrix of ξ, and $A(\gamma)$ is the matrix of partial derivatives of γ with respect to the parameters of the model (10.3.4). One can easily write $A(\gamma) = -\Delta^{-1}A^*(\gamma)$, where $A^*(\gamma)$ is given by

$$A^*(\gamma) = \begin{bmatrix}
0 & \frac{1}{2} & 0 & \ldots 0 & \gamma & 0 & \ldots 0 & \frac{\gamma_2}{2} & \frac{\gamma_3}{2} & \ldots 0\\
0 & 0 & \frac{1}{2} & \ldots 0 & 0 & \gamma_2 & \ldots 0 & \frac{\gamma_1}{2} & 0 & \ldots 0\\
\ldots & \ldots & \ldots & \ldots & \ldots & \ldots & \ldots & \ldots & \ldots & \ldots\\
0 & 0 & 0 & \ldots 0 & 0 & 0 & \ldots 0 & 0 & 0 & \ldots \frac{\gamma_{q-1}}{2}\\
0 & 0 & 0 & \ldots \frac{1}{2} & 0 & 0 & \ldots \gamma_{q-1} & 0 & 0 & \ldots \frac{\gamma_{q-2}}{2}
\end{bmatrix}.$$

The D-optimality criterion selects the optimum design by minimizing the determinant of $D(\hat{\gamma}) = [A^*(\gamma)M^{-1}(\xi)A^{*\prime}(\gamma)]$. However, the determinant depends on the unknown model parameters through γ, and, to remove the nuisance parameters, Pal and Mandal (2012c) used the pseudo-Bayesian approach of Pal and Mandal (2006). The second-order prior moments of $\boldsymbol{x}_0$ are assumed to be known, and since no prior information is available on $\boldsymbol{x}_0$, one may take $\mathcal{E}(x_{i0}^2)$ to be same for all $i = 1, 2, \ldots, q$ and $\mathcal{E}(x_{i0}x_{j0})$ to be equal for all $1 \leq i < j \leq q$. Since $\sum_{i=1}^{q} x_{i0} = 1$, this assumption is equivalent to stating that

$$\mathcal{E}(\gamma_i) = 0, \mathcal{E}(\gamma_i^2) = v, \mathcal{E}(\gamma_i\gamma_j) = w, \forall\ 1 \leq i < j \leq q-1, \tag{10.3.5}$$

where $v \in (0, 1)$, $w \in (-1, 1)$, and $v > w$.

Then, invocation of the D-optimality criterion amounts to minimizing

$$\phi_D^*(\xi) = \text{Det.}\mathcal{E}[A^*(\gamma)M^{-1}(\xi)A^{*\prime}(\gamma)]. \tag{10.3.6}$$

or, equivalently, to maximizing $\phi_D^{**}(\xi) = -\log \phi_D^*(\xi)$.

The function $\phi_D^{**}(\xi)$ possesses two important properties, viz. invariance and concavity (cf. Mandal 1982). Thus, to find the optimum design for estimating γ, and hence $\boldsymbol{x}_0$, one can restrict to the subclass of invariant designs and, in particular, to the subclass $\mathcal{D}_{\mathcal{R}}$ of invariant designs given by

$$\mathcal{D}_{\mathcal{R}} = \{\eta^* \mid \eta^* = \alpha_0\eta_0 + \alpha_1\eta_1 + \cdots + \alpha_{q-1}\eta_{q-1}, 0 \leq \alpha_i \leq 1,$$
$$\text{for } i = 1, 2, \ldots, q-1, \sum_{i=1}^{q-1} \alpha_i = 1\},$$

where for each $i = 2, 3, \ldots, q-1$, the design η_i is defined as:

$$\eta_i = \left\{ \begin{array}{cc} t \longleftrightarrow (\overbrace{1, 1, \ldots, 1}^{i}, \overbrace{0, 0, \ldots, 0}^{q-i-1})' & t \longleftrightarrow (\overbrace{-1, -1, \ldots, -1}^{i}, \overbrace{0, 0, \ldots, 0}^{q-i-1})' \\ \frac{1}{2C(q-1,i)} & \frac{1}{2C(q-1,i)} \end{array} \right\}$$

Here, $\boldsymbol{t} \longleftrightarrow \boldsymbol{r}$ means $\boldsymbol{t} = \boldsymbol{P}\boldsymbol{r}$, where P is some $(q-1) \times (q-1)$ permutation matrix. Huang and Huang (2009) also restricted to this subclass for the problem of estimation of quadratic coefficients, η_i assigns equal masses to the vertices of the experimental domain $\mathcal{F}$, with i components equal to $+1$ or -1. For example, in a three-component experiment, η_1 assigns mass $1/4$ to each of the points (1, 0), (0, 1), $(-1, 0)$, and $(0, -1)$, while η_2 assigns mass 1/2 to each of the points (1, 1) and $(-1, -1)$. Clearly, the designs η_i, $1 \leq i \leq q-1$, are $\mathcal{R}$-invariant, i.e., invariant with respect to permutation of coordinates and sign changes.

After some algebraic manipulation, one gets $\phi_D^*(\eta^*) = (m-n)^{q-2}\{m+(q-2)n\}$, where m and n are functions of v, w, α_i, $0 \leq i \leq q-1$. It may be mentioned that α_i is the weight assigned to the design η_i. The D-optimal design within $\mathcal{D}_{\mathcal{R}}$ is then obtained by minimizing $\phi_D^*(\eta^*)$ with respect to α_i, $0 \le i \le q-1$, where $0 \le \alpha_i \le 1$ and $\sum \alpha_i = 1$. The optimality or otherwise of the design within the entire class is verified using Kiefer (1974) equivalence theorem, which in the present setup, reduces to the following :

Theorem 10.3.1 (Equivalence Theorem) *A necessary and sufficient condition for a design ξ to be D-optimal is that*

$$tr[\mathcal{E}\{A^*(\gamma)M^{-1}(\xi)f(\boldsymbol{t})f'(\boldsymbol{t})M^{-1}(\xi)A^{*\prime}(\gamma)\}(\mathcal{E}\{A^*(\gamma)M^{-1}(\xi)A^{*\prime}(\gamma)\})^{-1}] \le q-1 \tag{10.3.7}$$

holds for all $\boldsymbol{t}$ in the factor space $\mathcal{F}$. Equality in (10.3.7) holds at the support points of ξ.

To express the optimal design in the experimental region S_δ^{q-1}, for each $\boldsymbol{x} \in S^{q-1}$, we use the notation $\boldsymbol{x} \sim (k_1, k_2, \ldots, k_q)$ to mean

$$\boldsymbol{x} = \frac{(k_1, k_2, \ldots, k_q)'}{\|(K-1, k_2, \ldots, k_q)\|},$$

where $\delta \le \frac{k_i}{k_j} \le 1/\delta, 1 \le i, j \le q, \delta \in (0,1)$ and $\| \|$ denotes the usual L_1 norm. Further, we write $\boldsymbol{x} \overset{q-1}{\longleftrightarrow} (k_1, k_2, \ldots, k_q)$ to denote the collection of all $\boldsymbol{x} \sim (k_{P(1)}, k_{P(2)}, \ldots, k_{P(q-1)}, k_q)$, for all permutation P of $\{1, 2, \ldots, q-1\}$. For example, if $q = 3$, then $\boldsymbol{x} \sim ((1, \delta, 1)$ means $\boldsymbol{x} = (1/(\delta+2), \delta/(\delta+2), 1/(\delta+2))'$; $\boldsymbol{x} \overset{2}{\longleftrightarrow} (1, \delta, 1)$ means $\boldsymbol{x} \in \{\boldsymbol{x} \in S^2 : \boldsymbol{x} \sim (1, \delta, 1)$ or $\boldsymbol{x} \sim (\delta, 1, 1)\}$. Then, the design η_i on the experimental region $\mathcal{F}$ corresponds to the design ξ_i on the experimental domain S_δ^{q-1} given by (10.3.1), where

$$\xi_i = \left\{ \begin{array}{cc} \boldsymbol{x} \overset{q-1}{\longleftrightarrow} (\overbrace{1, 1, \ldots, 1}^{i}, \overbrace{\delta, \delta, \ldots, \delta, \delta}^{q-i-1})' & \boldsymbol{x} \overset{q-1}{\longleftrightarrow} (\overbrace{\delta, \delta, \ldots, \delta}^{q-i-1}, 1, 1, \ldots, 1, 1)' \\ \frac{1}{2C(q-1,i)} & \frac{1}{2C(q-1,i)} \end{array} \right\}$$

For the case of $q = 2$, it can be shown that the D-optimal design assigns mass $\alpha_0/2$ at each of the support points $\left(\frac{\delta}{1+\delta}, \frac{1}{1+\delta}\right)$ and $\left(\frac{1}{1+\delta}, \frac{\delta}{1+\delta}\right)$, and mass $1-\alpha_0$ at the centroid $(1/2, 1/2)$, where $\alpha_0 = \frac{\sqrt{v+1/4}}{\sqrt{v}+\sqrt{v+1/4}}$.

Owing to the presence of apriori moments (v, w), verification of optimality of a design by equivalence theorem becomes algebraically rather involved for $q \ge 3$.

Table 10.3 Optimal masses assigned to η_is for some combinations of (v, w)

q	v	w	α_0	α_1	α_2	α_3	α_4
3	0.1	0.05	0.2265	0.5157	0.2578	–	–
	0.2	0.10	0.2572	0.4952	0.2476	–	–
	0.3	0.15	0.2712	0.4859	0.2429	–	–
	0.4	0.15	0.2805	0.4874	0.2321	–	–
	0.6	0.23	0.2483	0.5063	0.2454	–	–
4	0.1	0.07	0.1327	0.3747	0.3624	0.1302	–
	0.2	0.10	0.1630	0.3870	0.3210	0.1290	–
	0.4	0.15	0.1782	0.3934	0.3058	0.1226	–
	0.6	0.40	0.1787	0.3775	0.3061	0.1377	–
	0.8	0.25	0.1857	0.3966	0.2995	0.1182	–
5	0.1	0.05	0.0897	0.2942	0.3255	0.2170	0.0736
	0.3	0.10	0.1102	0.3153	0.3063	0.1966	0.0716
	0.4	0.20	0.1116	0.3111	0.2997	0.1998	0.0778
	0.6	0.40	0.1124	0.3037	0.2937	0.2061	0.0841
	0.8	0.25	0.1173	0.3201	0.3004	0.1914	0.0708

However, for $q = 3, 4, 5,$ Pal and Mandal (2012c) numerically verified, using innumerable points in the experimental domain, that the condition (10.3.7) is satisfied. Table 10.3 gives the D-optimal designs for $q = 3, 4, 5,$ for some combinations of (v, w), as obtained by Pal and Mandal (2012c).

References

Aitchison, J., & Bacon-Shone, J. (1984). Log contrast models for experiments with mixtures. *Biometrika, 71*, 323–330.

Atwood, C. L. (1969). Optimal and efficient designs of experiments. *The Annals of Mathematical Statistics, 40*, 1570–1602.

Chan, L. Y. (1992). D-optimal design for a quadratic log contrast model for experiments with mixtures. *Communications in Statistics-Theory and Methods, 20*, 2909–2930.

Chan, L. Y., Guan, Y. N., & Zhang, C. (1998a). A-optimal designs for an additive quadratic mixtures model. *Statistica Sinica, 8*, 979–990.

Chan, L. Y. (2000). Optimal designs for experiments with mixtures: a survey. *Communications in Statistics-Theory and Methods, 29*, 2281–2312.

Darroch, J. N., & Waller, J. (1985). Additivity and interaction in three-component experiments with mixtures. *Biometrika, 72*, 153–163.

Huang, M. K., & Huang, M. N. L. (2009). D_s-optimality designs for quadratic log contrast model for experiments with mixtures. *Communications in Statistics-Theory and Methods, 38*, 1607–1621.

Kiefer, J. (1974). General equivalence theory for optimum designs (approximate theory). *The annals of Statistics, 2*, 849–879.

Mandal, N. K. (1982). D-optimal designs for estimating the optimum point in a multifactor experiment. *Bulletin of the Calcutta Statistical Association, 31*, 105–130.

Mandal, N. K., Pal, M., & Aggarwal, M. L. (2012). Pseudo Bayesian A-optimal designs for estimating the point of maximum in component-amount Darroch Waller mixture model. *Statistics & Probability Letters*, *82*, 1088–1094.

Pal, M. Mandal, N. K., & Aggarwal, M. L. (2012). A-optimal designs for optimum mixture in an additive quadratic mixture model. Communicated.

Pal, M., & Mandal, N. K. (2006). Optimum designs for optimum mixtures. *Statistics & Probability Letters*, *76*, 1369–1379.

Pal, M., & Mandal, N. K. (2012a). Optimum designs for estimation of parameters in a quadratic mixture-amount model. *Communications in Statistics-Theory and Methods*, *41*, 665–673.

Pal, M., & Mandal, N. K. (2012b). D-optimal designs for an additive quadratic mixture model with random regression coefficients. *Journal of the Indian Society of Agricultural Statistics*, *66*, 441–446.

Pal, M., & Mandal, N. K. (2012c). D-optimum designs for optimum mixture in a quadratic log contrast model. *Communications in Statistics-Simulation and Computation*, *41*, 253–264.

Scheffé, H. (1958). Experiments with mixtures. Journal of the Royal Statistical Society. *Series B (Methodological)*, *20*, 344–360.

Zhang, C. Q., Chan, L. Y., Guan, Y. N., Li, K. H., & Lau, T. S. (2005). Optimum designs for an additive quadratic mixture model involving the amount of mixture. *Statistica Sinica*, *15*, 165–176.

Chapter 11
Applications of Mixture Experiments

Abstract The purpose of this chapter is to describe some application areas of mixture experiments. We present some studies taken up in the context of agricultural/horticultural/pharmaceutical experiments in the form of mixture designs in order to extract meaningful information for specific items of enquiry.

Keywords Applications of mixture experiments · Intercropping experiment · Ready-to-serve fruit beverage experiment · Pharmaceutical experiment

11.1 Introduction

Experimental designs have vast applications in different fields such as agriculture, engineering, pharmacy, biomedical, and environmental studies, to name a few. In such areas, often experiments are conducted with a fixed quantity of inputs (same dose of fertilizer, same quantity of irrigation water or same dose of diluent, etc.) applied as mixture of two or more components. Here, the 'treatments' are either combinations of two or more ingredients whose total is a fixed quantity or are quantities of inputs applied at different experimental stages such that the sum total of the quantities is fixed. These experiments are generally carried out using a randomized complete block design (RCBD) or a completely randomized design (CRD), and aim at identifying the best among the 'treatments' tried in the experiment. However, if the experimenter is also interested in obtaining a functional relationship between the proportions of inputs applied and the mean response, so that one can interpolate the responses at points that have not been tried in the experiment, and/or obtain the optimum proportion of the inputs, RCBD/CRD analysis alone is not sufficient. In this case, one has to draw an analogy of these experiments with mixture experiments. Many such experiments are being conducted in the National Agricultural Research System in India and elsewhere.

B. K. Sinha et al., *Optimal Mixture Experiments*, Lecture Notes in Statistics 1028,
DOI: 10.1007/978-81-322-1786-2_11,

Again, in studies related to ready-to-serve (RTS) beverages, suitable mixtures of the ingredients in appropriate proportions are to be ascertained, keeping an objective in mind. This naturally calls for a study in the setup of a mixture experiment. In pharmaceutical experiments, it is a natural problem to combine different excipients in an 'optimal' manner to 'maximize' the 'output' in some sense. In all of the above three areas, we will narrate examples of actual studies carried out as mixture experiments.

11.2 Application in Replacement Series Intercropping Experiment

Intercropping is an important feature of dryland farming and it has proved very useful for survival of small and marginal farmers in tropical and subtropical regions.

Depending upon the row arrangements, the intercropping experiments are classified into two types viz. (i) **additive series**, where the component crop is introduced without reducing the plant population of the main crop, and (ii) **replacement series**, where the component crop is introduced by replacing a part of the main crop. In the latter case, the 'treatments' are the different row ratios. A critical look at the treatment structure reveals that in such experiments, the total land resources, i.e., area under each experimental unit is constant and the response varies only due to different proportions of the crops. Thus, the variation in response is due to the varying proportions of the area allotted to the crops. The 'sole crop' treatments are the 'pure' blends, and the different proportions (treatments) are treated as 'mixtures.'

Let there be N design points (experimental plots) and v distinct intercropping treatment combinations (ITCs). In this situation, we assume $N = vr$ so that each ITC may be replicated r times. Let the proportionate area allocated to ith ITC in uth experimental unit be x_{iu}, $1 \leq i \leq v$; $1 \leq u \leq N$. The response is assumed to have functional relationship with the proportions allocated to the ITCs in the uth experimental unit, and the relationship may be explained by second-order canonical polynomial of Scheffé (1958). For two-component ITCs, this is given by

$$y_u = \beta_1 x_{1u} + \beta_2 x_{2u} + \beta_{12} x_{1u} x_{2u} + e_u,$$

where β_1, β_2 and β_{12} are the usual regression coefficients. Also,

$$x_{1u} + x_{2u} = 1, 0 \leq x_{1u}, x_{2u} \leq 1; \quad u = 1, 2, \ldots, N.$$

The parameters can be estimated using the ordinary least squares method, and thereby the estimates of the optimum mixing proportions of area allocated to different ITCs for maximizing the gross returns and the optimum response can be obtained.

Empirical Illustration: An intercropping experiment was conducted by Dr. Panjabrao Deshmukh Krishi Vidyapeeth, Dryland Research Station, Akola, India on Redgram and Safflower with the objective of finding the optimum proportions of area to be allocated to redgram and safflower so as to maximize the gross returns. There are five ITCs defined as

ITC_1: Sole Redgram
ITC_2: Redgram + Safflower in the row ratio of 2:1
ITC_3: Redgram + Safflower in the row ratio of 1:2
ITC_4: Redgram + Safflower in the row ratio of 1:1
ITC_5: Sole Safflower

Accordingly, the proportions of area allocated to crops redgram and safflower are 1:0, 2:1, 1:2, 1:1, 0:1.

For details, we refer to Dhekale (2001) and Dhekale et al. (2003).

The findings of the above study, as an application of mixture experiment, suggested the following 'optimum' proportion(s) of redgram for different price ratios [price ratio, x_1(opt)]:

$$(1{:}0.75, 24.49); (1{:}0.80, 20.20); (1{:}1.00, 11.15); (1{:}1.05, 9.86);$$
$$(1{:}1.10, 8.78); (1{:}1.30, 5.76); (1{:}1.50, 3.91); (1{:}1.75, 2.41).$$

11.3 Preparation and Standardization of RTS Fruit Beverages

An experiment was conducted at Division of Fruits and Horticultural Technology, Indian Agricultural Research Institute, New Delhi, India to study the feasibility of blending fruit juice/pulp of lime, aonla, grape, pineapple, and mango in different proportions (5–95 %) for preparation and standardization of ready-to-serve (RTS) beverages for improving the aroma, taste, and nutrients of the beverages. Four different combinations of fruit juices viz. lime–aonla, mango–pineapple, grape–pineapple, and mango–grape mixed in the ratios 0:100, 5:95, 10:90, 15:85, 20:80, 25:75, 50:50, 75:25, 80:20, 85:15, 90:10, 95:5, 100:0 were considered, and a panel of nine members, adopting nine-point hedonic scale organoleptically, evaluated the prepared beverages. The experiment was replicated three times, and the data were analyzed as a one-way classified data separately for all the four mixtures. For details, we refer to Deka et al. (2001). We highlight below some of the findings.

Since the age of the panel members (respondents) varied widely, which could have an effect on their perception of the beverages,the data were then divided into three groups according to the ages of the respondents, viz., 22–34, 35–44, and 45–55 years, and were then analyzed as a two-way classified data. Significant differences were observed among the age groups.

As the optimum blending proportion that maximizes the responses (viz. *hedonic scores on color, aroma, taste, and overall*) may be other than the proportions tried in the experiment, it was necessary to establish a relationship between each response and the mixing proportions. Scheffé's quadratic mixture model was used for the purpose. The data were then analyzed using mixtures methodology to estimate the parameters of the models.

For details, we refer to Deka et al. (2001). The following observations emerged from their data analysis:

(i) In lime–aonla mixture, sensory score decreases with decrease in lime proportion.
(ii) In mango–pineapple mixture, sensory score increases with decrease in mango proportion.
(iii) In grape–pineapple mixture, sensory score decreases with decrease in pineapple proportion.
(iv) In mango–grape mixture, there is not much change in sensory score with change in the mixture combination.

11.4 Application in Pharmaceutical Experiments

Pharmaceutical formulations are basically mixtures of a number of excipients. The main aim in such formulations is to study the effects of different proportions of the excipients on the characteristics of the formulation. These characteristics include both the final properties of the dosage form and the case of processing, which depend on the relative proportions of the constituents.

In almost all pharmaceutical formulations, there are constraints on the proportions of the constituents used. There can be two types of constraints, viz. absolute constraints and relational constraints. In an absolute constraint, the proportion of an excipient is governed by a lower and an upper bound, while in a relational constraint, lower and/or upper bounds are specified on a linear combination of the mixture combination. Relational constraints generally arise when the mixture contains constituents of the same type. For example, in an inert matrix tablet, besides other ingredients, there may be three types of polymers and two types of diluents. The polymers and diluents may then be classified into two groups under the broad heading 'polymer' and 'diluent,' and these are called the 'major' components or '*M*' components. The components within a group are referred to as 'minor' components or '*m*' components. The number of members of a *M* component is normally not very large, say two or three. (Cf. Lewis et al. 1998). Bounds on the proportions of the *M* components give relational constraints on the minor components.

Initially, Scheffé's first-order model was used to define the mean response in a pharmaceutical experiment with two major components and relational constraints. However, owing to the different permeabilities of the polymers, their relative proportions were expected to affect the dissolution properties of the tablet. Hence, a second-order term indicating the interaction between the polymers was added. Further extensions assumed interaction between polymers and diluents.

In the special case, where the proportions of the major components are fixed, a product model was found to be more appropriate, and it was meant to define mean response as a second-order Scheffé model in polymers, but each coefficient of the model depended on the composition of the diluents according to a first-order Scheffé model. Thus, for two major components—polymer (A and B) and diluent (D_1, D_2, D_3)—the response is given by

$$\begin{aligned}\eta_x &= \beta_1 x_1 + \beta_2 x_2 + \beta_{12} x_1 x_2 \\ &= (\theta_{13} x_3 + \theta_{14} x_4 + \theta_{15} x_5) x_1 + (\theta_{23} x_3 + \theta_{24} x_4 + \theta_{55} x_5) x_2 \\ &\quad + (\theta_{123} x_3 + \theta_{124} x_4 + \theta_{125} x_5) x_1 x_2. \end{aligned} \tag{11.4.1}$$

where (x_1, x_2) denotes the proportions of polymers A and B and (x_3, x_4, x_5) denotes the proportions of diluents D_1, D_2, and D_3 in the mixture. Not much research has been done for the above type of response function. The natural questions that arise here are (i) what should be the parameters of the model? and (ii) what is the optimum polymer composition? Lewis et al. (1998) indicated that the *D*-optimal design for parameter estimation is a nine-point design obtained by multiplying the second-order Scheffé design for the polymers (3 points) and the first-order Scheffé design for diluents (3 points). However, investigation for obtaining the optimal design for estimating the optimum proportions of the polymers has rarely been made.

Pal and Mandal (2013) considered model (11.4.1) with two major components, M_1 and M_2, having m and n minor components, respectively, and occurring in fixed proportions in the mixture. The response function is quadratic in the minor components of M_1, and each regression coefficient is linear in the minor components of M_2.

Let $(x_1, x_2, \ldots, x_m)$ and $(x_{m+1}, x_{m+2}, \ldots, x_{m+n})$ denote the proportions of the minor components in M_1 and M_2, respectively, where $0 \leq x_i \leq 1$, $\sum_{i=1}^{m+n} x_i = 1$. Suppose the proportions of the major components are fixed as $\sum_{i=1}^{m} x_i = \delta$, $\sum_{i=m+1}^{m+n} x_i = 1 - \delta, 0 < \delta < 1$.

The mean response $\eta(\boldsymbol{x})$ is given by

$$\eta(\boldsymbol{x}) = \sum_{i=1}^{m} \beta_i x_i + \sum_{i<j=1}^{m} \beta_{ij} x_i x_j,$$

where

$$\beta_i = \sum_{k=m+1}^{m+n} \theta_{ik} x_k, \quad i = 1, 2, \ldots, m, \quad \beta_{ij} = \sum_{k=m+1}^{m+n} \theta_{ijk} x_k, \quad i < j = 1, 2, \ldots, m,$$

and the experimental region is given by

$$\Xi = \{(x_1, x_2, \ldots, x_{m+n}) \mid 0 \leq x_i \leq 1, \quad i = 1, 2, \ldots, m+n, \quad \sum_{i=1}^{m} x_i = \delta, \sum_{i=1}^{m+n} x_i = 1\}. \tag{11.4.2}$$

One can express the experimental region as $\Xi = \Xi_1 \cap \Xi_2$, where

$$\Xi_1 = \left\{ (x_1, x_2, \ldots, x_m) \mid 0 \leq x_i \leq 1, \quad i = 1, 2, \ldots, m, \sum_{i=1}^{m} x_i = \delta \right\} \tag{11.4.3}$$

$$\Xi_2 = \left\{ (x_{m+1}, x_{m+2}, \ldots, x_{m+n}) \mid 0 \le x_i \le 1, \quad i = m+1, \ldots, m+n, \ \sum_{i=m+1}^{m+n} x_i = 1 - \delta \right\} \tag{11.4.4}$$

Henceforth, we shall write $\boldsymbol{x}_{(1)} = (x_1, x_2, \ldots, x_m)$ and $\boldsymbol{x}_{(2)} = (x_{m+1}, x_{m+2}, \ldots, x_{m+n})$.

11.4.1 Optimum Design for Parameter Estimation

Consider the class $\mathcal{D}$ of all competing continuous designs, for which all the parameters of (11.4.2) are estimable. We want to find a continuous design ξ in $\mathcal{D}$ that can estimate the parameters with maximum accuracy.
Pal and Mandal (2013) proved the following theorem :

Theorem 11.4.1 *The D-optimal (as also trace-optimal) design contains support points which are union of barycentres of Ξ_1 and Ξ_2.*

***D*-optimality criterion**
In view of Theorem 11.4.1, let us first consider a subclass $\mathcal{D}_{12}$ of designs ξ with support points given by

(i) $(\delta, 0, \ldots, 0; 1-\delta, 0, \ldots, 0)$ and all possible permutations within the first m coordinates and within the last n coordinates, respectively, each with weight $\alpha_1 = \alpha/mn$—having equal split of total weight α assigned to these support points;
(ii) $(\delta/2, \delta/2, 0, \ldots, 0; 1-\delta, 0, \ldots, 0)$ and all possible permutations within the first m coordinates and within the last n coordinates, respectively, each with weight $\alpha_2 = (1-\alpha)/nC(m, 2)$.

Thus, whereas $0 < \alpha < 1$ and it splits into mn equal parts, $(1-\alpha)$ splits itself equally among $nC(m, 2)$ terms. We shall refer to the support points in (i) as pure-type support points and those in (ii) as mixed-type support points.

For any design $\xi \in \mathcal{D}_{12}$, the information matrix is given by

$$M(\xi) = X \Lambda X',$$

where

$$X = \begin{bmatrix} \delta(1-\delta) I_m & \frac{\delta(1-\delta)}{2} P \\ 0 & \frac{\delta^2(1-\delta)}{4} I_{C(m,2)} \end{bmatrix} \otimes I_n,$$

$$P^{m\times C(m,2)} = \begin{bmatrix} \overbrace{1\,1\ldots1}^{m-1} & \overbrace{0\,0\ldots0}^{m-2} & \overbrace{0\,0\ldots0}^{m-3} & \ldots & \overbrace{0}^{1} \\ 1\,0\ldots0 & 1\,1\ldots1 & 0\,0\ldots0 & \ldots & 0 \\ \ldots\ldots & \ldots\ldots & \ldots\ldots & \ldots & \\ 0\,0\ldots1 & 0\,0\ldots1 & 0\,0\ldots1 & \ldots & 1 \end{bmatrix}$$

$$\Lambda = \begin{bmatrix} \alpha_1 I_m & 0 \\ 0 & \alpha_2 I_{C(m,2)} \end{bmatrix} \otimes I_n.$$

Therefore,

$$M^{-1}(\xi) = \frac{n}{\delta^2(1-\delta)^2} M_1^{-1}(\xi) \otimes I_n,$$

where

$$M_1^{-1}(\xi) = \begin{bmatrix} \frac{1}{\alpha_1^*} I_m & -\frac{2}{\delta\alpha_1^*} P \\ -\frac{2}{\delta\alpha_1^*} P' & \frac{4}{\delta^2\alpha_1} P'P + -\frac{16}{\delta^2\alpha_2^*} I_{C(m,2)} \end{bmatrix}, \tag{11.4.5}$$

with

$$\alpha_1^* = n\alpha_1, \quad \alpha_2^* = n\alpha_2.$$

Since the design ξ is saturated, the D-optimal design, say ξ_D, in $\mathcal{D}_{12}$ has $\alpha = mn/p$. Using equivalence theorem, Pal and Mandal (2013) established the following :

Theorem 11.4.2 *The D-optimal design for estimation of parameters in model* (11.4.2) *is a saturated design with pure- and mixed-type support points.*

Now we turn to the A-optimality criterion and present relevant optimality results.

For trace optimality, the criterion function is $\phi(\xi) = Trace(M^{-1}(\xi))$. Within the subclass $\mathcal{D}_{12}$, the trace-optimal design has

$$\alpha = \alpha_0 = \frac{\sqrt{a}}{\sqrt{a}+\sqrt{b}}, \tag{11.4.6}$$

where

$$a = m(m + 8C(m,2))/\delta^2; b = 16C(m,2)^2/\delta^2.$$

Pal and Mandal (2013) proved the following:

Theorem 11.4.3 *For $m \neq 3$, the trace-optimal design for estimation of parameters in model* (11.4.2) *has support points of the pure type, each with weight $\frac{\alpha_0}{mn}$, and the mixed type, each with weight $\frac{1-\alpha_0}{nC(m,2)}$, where α_0 is given by* (11.4.6).

Theorem 11.4.4 *When $m = 3$, the trace-optimal design, for estimation of parameters in the mixture model* (11.4.2) *has pure-type support points, each with weight α_1^*, mixed-type support points, each with weight α_2^*, and support points of the type $(\delta/3, \delta/3, \delta/3; 1-\delta, 0, \ldots, 0)$, and all possible permutations within the last n coordinates, each with weight α_3^*, where $\alpha_1^* = 0.1417$, $\alpha_2^* = 0.1873$, $\alpha_3^* = 0.0130$.*

11.4.2 Optimum Design for Estimation of the Optimum Composition of M_1

Using the constraint $\sum_{i=1}^{m} x_i = \delta$, we can re-write the mean response function in (11.4.2) as

$$\eta(\boldsymbol{x}) = \sum_{i=1}^{m} \lambda_{ii} x_i^2 + \sum_{i<j=1}^{m} \lambda_{ij} x_i x_j = \boldsymbol{x}'_{(1)} \boldsymbol{B} \boldsymbol{x}_{(1)}, \tag{11.4.7}$$

where

$$\begin{aligned}
\lambda_{ii} &= \frac{\beta_i}{\delta}, \quad \lambda_{ij} = \frac{\beta_i}{\delta} + \frac{\beta_j}{\delta} + \beta_{ij}, \quad B = ((b_{ij})),\\
b_{ij} &= \frac{1}{2}(1 + \delta_{ij})\lambda_{ij}, \quad i, j = 1, 2, \ldots, m,\\
\delta_{ij} &= 1, \quad \text{if } i = j\\
&= 0, \quad \text{if } i \neq j.
\end{aligned}$$

Clearly, λ_{ij}s are linear functions of $x_{m+1}, x_{m+2}, \ldots, x_{m+n}$. Let,

$$\lambda_{ij} = \sum_{k=m+1}^{m+n} \mu_{ijk} x_k. \tag{11.4.8}$$

From (11.4.7), the optimum proportions in M_1 for given $\boldsymbol{x}_{(2)}$ is $\gamma = \delta(1'B^{-1}1)^{-1} B^{-1}1 = \rho B^{-1}1$, where $\rho = \delta(1'B^{-1}1)^{-1}$.

For a design ξ, the large sample conditional dispersion matrix of $\hat{\gamma}$, given $\boldsymbol{x}_{(2)}$, is

$$D(\hat{\gamma} \mid \boldsymbol{x}_{(2)}) = A(\gamma) M^{-1}(\xi) A'(\gamma),$$

where

$$\begin{aligned}
A(\gamma) &= \left(\frac{\partial\gamma}{\partial\mu_{111}}, \ldots, \frac{\partial\gamma}{\partial\mu_{11n}}, \ldots, \frac{\partial\gamma}{\partial\mu_{mm1}}, \ldots, \frac{\partial\gamma}{\partial\mu_{mmn}}, \ldots, \frac{\partial\gamma}{\partial\mu_{121}}, \ldots, \frac{\partial\gamma}{\partial\mu_{m-1,m,n}} \right)\\
&= A^*(\gamma) C(\boldsymbol{x}_{(2)}), \text{ say,}
\end{aligned}$$

with

$$\begin{aligned}
A^*(\gamma) &= \left(\frac{\partial\gamma}{\partial\lambda_{11}}, \ldots, \frac{\partial\gamma}{\partial\lambda_{11}}, \ldots, \frac{\partial\gamma}{\partial\lambda_{mm}}, \ldots, \frac{\partial\gamma}{\partial\lambda_{mm}}, \frac{\partial\gamma}{\partial\lambda_{12}}, \ldots, \frac{\partial\gamma}{\partial\lambda_{12}}, \ldots, \frac{\partial\gamma}{\partial\lambda_{m-1,m}} \right)\\
&= d(1'_n \otimes a_1, \ldots, 1'_n \otimes a_{m+C(m,2)})
\end{aligned}$$

being an $m \times p$ matrix and a_is being the columns of the matrix

$$A_1(\gamma) = d \begin{pmatrix} -2(m-1)\gamma_1 2\gamma_2 & \cdots & 2\gamma_m\gamma_1 - (m-1)\gamma_2 & \cdots & \gamma_{m-1} + \gamma_m \\ 2\gamma_1 - 2(m-1)\gamma_2 & \cdots & 2\gamma_m\gamma_2 - (m-1)\gamma_1 & \cdots & \gamma_{m-1} + \gamma_m \\ 2\gamma_1 2\gamma_2 & \cdots & 2\gamma_m\gamma_1 + \gamma_2 & \cdots & \gamma_{m-1} + \gamma_m \\ \cdots\cdots\cdots & \cdots & \cdots\cdots\cdots & \cdots & \cdots\cdots\cdots \\ 2\gamma_1 2\gamma_2 & \cdots & 2\gamma_m\gamma_1 + \gamma_2 & \cdots & \gamma_{m-1} + \gamma_m \\ \cdots\cdots\cdots & \cdots & \cdots\cdots\cdots & \cdots & \cdots\cdots\cdots \\ 2\gamma_1 2\gamma_2 & \cdots & 2\gamma_m\gamma_1 + \gamma_2 & \cdots & \gamma_{m-1} - (m-1)\gamma_m \\ 2\gamma_1 2\gamma_2 & \cdots & -2(m-1)\gamma_m\gamma_1 + \gamma_2 & \cdots & \gamma_m - (m-1)\gamma_{m-1} \end{pmatrix}. \tag{11.4.9}$$

and d being a constant independent of the design (cf. Pal and Mandal (2006), and $C(\boldsymbol{x}_2) = Diag(x_{m+1}, \ldots, x_{m+n}, \ldots, x_{m+1}, \ldots, x_{m+n})$ is a $p \times p$ matrix.

However, $D(\hat{\gamma} \mid \boldsymbol{x}_{(2)})$ is a function of the unknown parameters of the model so that one may use the pseudo-Bayesian approach of Pal and Mandal (2006). Since the elements of $A^*(\gamma)'A^*(\gamma)$ are quadratic functions of γ_is, $i = 1, 2, \ldots, m$, we assume the prior moments of $\gamma_i\gamma_j$, $i, j = 1, 2, \ldots, m$ as follows :

$\mathcal{E}(\gamma_i^2 \mid \boldsymbol{x}_{(2)}) = v(\boldsymbol{x}_{(2)})$ for all $i = 1, 2, \ldots, m$, and $\mathcal{E}(\gamma_i\gamma_j \mid \boldsymbol{x}_{(2)}) = w(\boldsymbol{x}_{(2)})$, for every $i < j = 1, 2, \ldots, m$.
(Cf. Chap. 7.)
Since $\sum_{i=1}^m \gamma_i = \delta$, $v(\boldsymbol{x}_{(2)})$ and $w(\boldsymbol{x}_{(2)})$ satisfy $m[v(\boldsymbol{x}_{(2)}) + (m-1)w(\boldsymbol{x}_{(2)})] = \delta^2$.
Also, since $v(\boldsymbol{x}_{(2)}) \geq w(\boldsymbol{x}_{(2)})$, we have $\frac{\delta^2}{m^2} \leq v(\boldsymbol{x}_{(2)}) < \frac{\delta^2}{m}$, whatever be $\boldsymbol{x}_{(2)}$.

The trace criterion function is then given by

$$\begin{aligned}\phi(\xi, \boldsymbol{x}_{(2)}) &= \text{Trace}\,[M^{-1}(\xi)C(\boldsymbol{x}_{(2)})'\mathcal{E}\{A^*(\gamma)'A^*(\gamma)\}C(x_{(2)})]. \\ &= \text{Trace}\,[M^{-1}(\xi)(\mathcal{E}\{A_1(\gamma)'A_1(\gamma)\} \otimes (x_{(2)}x_{(2)}'))].\end{aligned}$$

To remove the effect of $\boldsymbol{x}_{(2)}$ from the criterion function, one may integrate $\phi(\xi, \boldsymbol{x}_{(2)})$ with respect to $\boldsymbol{x}_{(2)}$, and obtain the modified criterion as

$$\Phi(\xi) = \int\int \cdots \int \phi(\xi, \boldsymbol{x}_{(2)}) \prod_{j=1}^{n} dx_{m+j}.$$

$$\begin{gathered} 0 < x_{m+j} < 1 - \delta \\ 1 \leq j \leq n \\ \sum_{j=1}^{n} x_{m+j} = 1 - \delta \end{gathered}$$

Pal and Mandal (2013) considered the simplest situation where the components of M_2 do not affect the prior moments of γ, that is

$$v(\boldsymbol{x}_{(2)}) = v, \; w(\boldsymbol{x}_{(2)}) = w.$$

Then, ignoring the constant term,the criterion function becomes

$$\Phi_1(\xi) = \text{Trace}[M^{-1}(\xi)(\mathcal{E}\{A_1(\gamma)'A_1(\gamma)\} \otimes (I_n + 11'))], \tag{11.4.10}$$

Clearly, there is invariance in the criterion function with respect to the components of major components M_1 and M_2. Hence, we may restrict to the class of designs invariant with respect to the m components of M_1 and with respect to the n components of M_2.
Pal and Mandal (2013) established the following results :

Theorem 11.4.5 *The possible support points of the trace-optimal design for estimating γ are the union of barycentres of the component Ξ_1 and Ξ_2.*

Theorem 11.4.6 *For $2 \leq m \leq 4$ and given second-order prior moments of the optimum proportions γ in M_1, the trace-optimal design for estimating γ has the support points of pure type, each with weight $\frac{\alpha_0}{mn}$, and of mixed type, each with weight $\frac{1-\alpha_0}{nC(m,2)}$, where $\alpha_0 = \frac{\sqrt{a}}{\sqrt{a}+\sqrt{b}}$, with*

$$a = 2mn[v + (m-1)w][m + C(m,2)(m-2)]$$
$$b = 32nC(m,2)^2[(m-1)v - w].$$

Optimum replications at the support points of the optimum design in 100 runs of the experiment for estimation of γ, for some combinations of the prior moments (v, w), when $2 \leq m \leq 4$. Let r_1 denote the total number of replications of the pure-type support points and r_2 the total number of replications of the mixed-type support points. Within each type, the support points should be replicated as evenly as possible (Table 11.1).

Table 11.1 Optimum replications of the pure–and mixed-type support points of the trace-optimal designs for some combinations of (δ, v) in 100 runs of the experiment, when $2 \leq m \leq 4$

m	δ	v	r_1	r_2
2	0.4	0.05	50	50
		0.06	37	63
	0.6	0.10	60	40
		0.16	36	64
	0.8	0.20	50	50
		0.30	35	65
3	0.4	0.03	27	73
		0.05	21	79
	0.6	0.06	29	71
		0.10	22	78
	0.8	0.10	30	70
		0.15	24	76
4	0.4	0.016	25	75
		0.03	18	82
	0.6	0.04	24	76
		0.07	18	82
	0.8	0.06	26	74
		0.14	17	83

References

Deka, B. C., Sethi, V., Parsad, R., & Batra, P. K. (2001). Application of mixtures methodology for beverages from mixed fruit juice/pulp. *Journal of Food Science and Technology*, *38*, 615–618.

Dhekale, J. S. (2001). Analytical Techniques in Intercropping Experiments. Unpublished Ph. D. thesis, IARI, New Delhi.

Dhekale, J. S., Prasad, R., & Gupta, V. K. (2003). Analysis of intercropping experiments using experiments with mixtures methodology. *Journal of the Indian Society of Agricultural Statistics*, *56*, 260–266.

Lewis, G. A., Mathieu, D., & Phan-Tan-Luu, R. (1998). *Pharmaceutical experimental design*. New York: Marcel Dekker Inc.

Pal, M., & Mandal, N. K. (2006). Optimum designs for optimum mixtures. *Statistics & Probability Letters*, *76*, 1369–1379.

Pal, M., & Mandal, N. K. (2013). Optimum designs for mixtures with relational constraints on the components. *International Journal of Experimental Design and Process Optimisation*, *3*, 276–293.

Scheffé, H. (1958). Experiments with mixtures. Journal of the Royal Statistical Society. *Series B (Methodological)*, *20*, 344–360.

Chapter 12
Miscellaneous Topics: Robust Mixtures, Random Regression Coefficients, Multi-response Experiments, Mixture–Amount Models, Blocking in Mixture Designs

Abstract In this chapter, we dwell on some mixture design settings and present the underlying optimal designs. The purpose is to acquaint the readers with a variety of interesting and nonstandard areas of mixture designs. The chapter is divided into two parts. In Part A, we cover robust mixture designs and optimality in Scheffé and D–W models with random regression coefficients. In Part B, we discuss mixture–amount model due to Pal and Mandal (Comm Statist Theo Meth 41:665–673, 2012a), multi-response mixture models and mixture designs in blocks. We present the results already available and also some recent findings.

Keywords Robust mixture designs · Scheffé's quadratic mixture model · Darroch–Waller model · Random regression coefficients · Multi-response experiment · Mixture–amount models · Mixture designs in blocks · Optimal mixture designs

12.1 Robust Mixture Designs

12.1.1 Preliminaries

Optimum designs are concerned mainly with linear and/or non-linear function(s) of the parameters of the assumed model. Box and Draper (1959) first considered the problem of selecting a design when the assumed response function is inadequate in representing the true situation. They proposed the integrated mean square error (*IMSE*) criterion

$$\text{IM} = (N/\sigma^2)\int_{x} w(\boldsymbol{x})E\{\hat{y}(\boldsymbol{x}) - \eta(\boldsymbol{x})\}^2 \text{d}\boldsymbol{x} \qquad (12.1.1)$$

in selecting a design, where $\hat{y}(\boldsymbol{x})$ is the estimated response function, $\eta(\boldsymbol{x})$ is the true response function, N is the number of observations taken, σ^2 is the error variance and $w(\boldsymbol{x})$ is the mass attached to the point $\boldsymbol{x}$ in the region of interest $\mathcal{X}$ satisfying

B. K. Sinha et al., *Optimal Mixture Experiments*, Lecture Notes in Statistics 1028,
DOI: 10.1007/978-81-322-1786-2_12, © Springer India 2014

$$\int_{\mathcal{X}} w(\boldsymbol{x})\mathrm{d}\boldsymbol{x} = 1.$$

It is clear that IM given by (12.1.1) can be divided into two parts, viz.

$$\mathrm{IM} = V + B \tag{12.1.2}$$

where

$$V = (N/\sigma^2)\int_{\mathcal{X}} w(\boldsymbol{x})E\{\hat{y}(\boldsymbol{x}) - E(\hat{y}(\boldsymbol{x})\}^2\mathrm{d}\boldsymbol{x},$$

$$B = (N/\sigma^2)\int_{\mathcal{X}} w(\boldsymbol{x})E\{E(\hat{y}(\boldsymbol{x})) - \eta(\boldsymbol{x})\}^2\mathrm{d}\boldsymbol{x}. \tag{12.1.3}$$

In practice, it is difficult to find a design minimizing IM even in the simple setup. In the context of response surface design, Box and Draper (1959) observed that a design which minimizes IM is very close to a design which minimizes B, they called 'all-bias' design (also known as *BD* design) and may be quite different from one which minimizes V, the 'all-variance' design. They mainly restricted their investigations to lower-order polynomials, and the true response function and the assumed response function differ by degree one. While the support points of the V-optimum design, for the linear case, are mainly the extreme points of the factor space, those for the B-optimum design and the design with respect to the *IMSE*-criterion for many parameter combinations are the points inside the domain. Similar is the case when the response function is quadratic and the true response function is cubic.

Draper and Lawrence (1965a,b) considered the problem of finding designs in a mixture experiment with three and four factors. In situations, when the true response function is different from the assumed model, they compared designs minimizing B with other designs especially for the first-degree model with quadratic bias. However, for the quadratic model with cubic bias, they compared minimum bias design with the special family of designs obtained by scaling the *BD* design with no comparison with the D-optimal design. Galil and Kiefer (1977) have also considered this problem and showed that in some situations, depending on the values of the parameters in the bias term, the *BD* designs are inferior to the D-optimal design with respect to *IMSE*integrated mean square error (IMSE) criterion.

Since the *IMSE* criterion depends on unknown parameters, it is difficult to find designs minimizing (12.1.2). Chakrabarti and Mandal (1995) determined all-variance and all-bias designs and then compare them in the light of the *IMSE* criterion for some chosen values of unknown parameters.

For simplicity, we assume uniform weights throughout so that we can ignore $w(\boldsymbol{x})$ in the expression of IM and hence in B and V.

Box and Draper (1959) observed that the points inside the domain are informative with respect to the *IMSE* criterion. To find optimum designs minimizing *V*, *B* or *V* + *B*, Chakrabarti and Mandal (1995) considered a subclass of augmented axial designs which have the following three sets of points :

$$\begin{aligned}
x_{i,u} &= 1/q + \delta_1, x_{j,u} = 1/q - \delta_1/(q-1); \\
&\quad u = 1, 2, \ldots, q; \quad i = u; \quad j(\neq i) = 1, 2, \ldots, q; 0 \le \delta_1 \le (q-1)/q, \\
x_{i,u} &= 1/q - \delta_2, x_{j,u} = 1/q + \delta_2/(q-1); \\
&\quad u = q+1, \ldots, 2q; \quad i = u - q; \quad j(\neq i) = 1, 2, \ldots, q; 0 \le \delta_2 \le 1/q \\
x_{i,u} &= 1/2; x_{j,u} = 1/2; x_{k,u} = 0; \\
&\quad u = 2q+1, \ldots, q(q+3)/2, \quad k \neq (i < j) = 1, 2, \ldots, q. \qquad (12.1.4)
\end{aligned}$$

For $q = 3$, the design becomes an axial design (for axial and other designs see Cornell 2002). The weights attached with each point of the above three sets of design points are α/q, β/q, and $(1-\alpha-\beta)/C(q,2)$, respectively. The authors denoted such a subclass of designs by

$$\Delta_q = \{\Delta(\delta_1, \delta_2; \alpha, \beta; q); 0 \le \delta_1 \le (q-1)/q, 0 \le \delta_2 \le 1/q; 0 \le \alpha, \beta; \alpha + \beta \le 1\}. \qquad (12.1.5)$$

Let us write the assumed response function as $E(y/\boldsymbol{x}) = \boldsymbol{f}'_1(\boldsymbol{x})\theta_1$ while the true response function is given by $\eta_{\boldsymbol{x}} = \boldsymbol{f}'_1(\boldsymbol{x})\theta_1 + \boldsymbol{f}'_2(\boldsymbol{x})\theta_2$. Moreover, let us introduce the following notations of the moment matrices for a design ξ:

$$M_{ij}(\xi) = \int_R \boldsymbol{f}_i(\boldsymbol{x})\boldsymbol{f}'_j(\boldsymbol{x})\mathrm{d}\xi(\boldsymbol{x}); \quad i, j = 1, 2$$

and

$$M = ((M_{ij}, \quad i, j = 1, 2)).$$

Let the degree of the assumed polynomial response function be d_1 and that of the true response function be $d_1 + d_2$. When $d_1 = d_2 = 1$, $\boldsymbol{f}_1(\boldsymbol{x}), \boldsymbol{f}_2(\boldsymbol{x}), \boldsymbol{\theta}_1, \boldsymbol{\theta}_2$ have the form

$$\begin{aligned}
\boldsymbol{f}'_1(\boldsymbol{x}) &= (x_1, x_2, \ldots, x_k)', \qquad & \boldsymbol{f}'_2(\boldsymbol{x}) &= (x_1x_2, x_1x_3 \ldots, x_{k-1}x_k)', \\
\boldsymbol{\theta}_1 &= (\beta_1, \beta_2, \ldots, \beta_k)', \qquad & \boldsymbol{\theta}_2 &= (\beta_{12}, \beta_{13}, \ldots, \beta_{k-1,k})'.
\end{aligned}$$

Similarly, $\boldsymbol{f}_1(\boldsymbol{x}), \boldsymbol{f}_2(\boldsymbol{x}), \boldsymbol{\theta}_1, \boldsymbol{\theta}_2$ can be expressed in the same way for other values of d_1, d_2. Now, V is given by

$$V = \int_R \boldsymbol{f}'_1(\boldsymbol{x})M_{11}^{-1}\boldsymbol{f}_2(\boldsymbol{x})\mathrm{d}\boldsymbol{x}$$

$$= \text{Trace } M_{11}^{-1}\mu_{11}. \tag{12.1.6}$$

where $\mu_{11} = \int_R \boldsymbol{f}_1(\boldsymbol{x})\boldsymbol{f}'_1(\boldsymbol{x})\text{d}(\boldsymbol{x})$.

In general, let us denote μ_{ij} by

$$\mu_{ij} = \int_R \boldsymbol{f}_i(\boldsymbol{x})\boldsymbol{f}'_j(\boldsymbol{x})\text{d}(\boldsymbol{x}); \quad i, j = 1, 2. \tag{12.1.7}$$

Again, since $E(\hat{y}(\boldsymbol{x})) = \boldsymbol{f}'_1(\boldsymbol{x})[\boldsymbol{\theta}_1 + A\boldsymbol{\theta}_2]$, $A = M_{11}^{-1}M_{12}$, B takes the form

$$B = \int \boldsymbol{\theta}'_2[A'\boldsymbol{f}_1(\boldsymbol{x}) - \boldsymbol{f}_2(\boldsymbol{x})][A'\boldsymbol{f}_1(\boldsymbol{x}) - \boldsymbol{f}_2(\boldsymbol{x})]'\boldsymbol{\theta}_2\text{d}\boldsymbol{x}. \tag{12.1.8}$$

As in Box and Draper (1959), Chakrabarti and Mandal (1995) also considered the case $d_2 = d_1 + 1$. Moreover, they restricted to the cases $d_1 = 1, 2$.

12.1.2 V-Optimum Design

A V-optimum design minimizes $V = \text{Trace}M_{11}^{-1}\mu_{11}$. When $d_1 = 1$,

$$\mu_{11} = \frac{1}{q(q+1)}(I_q + J_{q,q}),$$

where I_q stands for the identity matrix of order q and $J_{q,q}$ stands for the $q \times q$ matrix with all elements unity. Utilizing of the convexity and invariance of the criterion function V with respect to the components, Chakrabarti and Mandal (1995) obtained the following result:

Theorem 12.1.1 *For $d_1 = 1$, the all-variance design puts equal mass at the vertices of a $(q, 1)$ simplex design.*

However, later, as mentioned in previous chapters, Draper and Pukelsheim (1999) and Draper et al. (2000) showed that given any invariant mixture design ξ with moment matrix $M(\xi)$, there exists a weighted centroid design (WCD) η with moment matrix $M(\eta)$, which is as good as $M(\xi)$ in the sense of partial Loewner order (PLO), i.e., $M(\eta) - M(\xi) \geq 0$. Hence, one can confine to the class of WCDs in finding a V-optimum design.

Remark 12.1.1 The results of Draper and Pukelsheim (1999) and Draper et al. (2000) are important in the sense that for any optimality criterion based on information matrix, if the criterion function is invariant, one may confine to the class of WCDs. Because of the above observations, to find an all-variance design, we may restrict ourselves to the class of WCDs.

Table 12.1 Computed values of the parameters

q	α	β	δ_1	δ_2
3	0.5	0.5	0.4137	0.2360
4	0.5	0.5	0.4111	0.2366
5	0.3	0.5	0.4017	0.1432
6	0.3	0.5	0.3296	0.1488

It is to be noted that Chakrabarti and Mandal (1995) arrived at the same support points directly without using the result of Draper and Pukelsheim (1999) .

For $d_1 = 2$, it is difficult to find a *V*-optimum design. As observed in Chap. 4, $a(q, 2)$ simplex design is *D*-optimum. Using equivalence theorem, Chakrabarti and Mandal (1995) showed that the *D*-optimum design in the whole class is not *V*-optimum. Instead, they found numerically the *V*-optimum design in the subclass Δ_q, given by (12.1.5), and compared the performance of their design with that of the *D*-optimum design for $q = 3, 4, 5$. The authors observed that the *V*-optimum design in Δ_q is better than the *D*-optimum design in respect of all-variance optimality criterion.

However, since the *V*-optimality criterion given by (12.1.6) is invariant with respect to the components, using Draper et al. (2000) , one can concentrate on the subclass of *WCD*s. Then, one needs to compute the masses attached to the different elementary centroid designs η_js to find the *V*-optimum design.

12.1.3 Minimum Bias Design

To find a minimum bias design, we have to minimize B given by (12.1.8). It is known that the minimum of (12.1.8) is attained at

$$M_{11} = \mu_{11}, \quad M_{12} = \mu_{12}, \tag{12.1.9}$$

and the minimum bias is then given by

$$B_{\min} = \boldsymbol{\theta}_2'(\mu_{22} - \mu_{12}'\mu_{11}^{-1}\mu_{12})\boldsymbol{\theta}_2. \tag{12.1.10}$$

Chakrabarti and Mandal (1995) computed the values of the parameters of the design in the subclass Δ_q satisfying (12.1.9) when $(d_1, d_2) = (1, 1), (2, 1)$ for $q = 3, 4, 5, 6$. These are reproduced below in Tables 12.1 and 12.2.

It is already mentioned that it is difficult to find designs minimizing the *IMSE* criterion (12.1.1). It involves unknown parameters of the model. Moreover, it does not have the well-known properties generally possessed by the classical optimality criteria. Box and Draper (1959) compared the performance of the all-variance and all-bias designs in respect of the *IMSE* criterion. It is quite obvious that the optimum *IMSE* design would depend heavily on whether $\boldsymbol{\theta}_2/\sigma^2$ is large or small. However,

Table 12.2 Computed values of the parameters

q	α	β	δ_1	δ_2
3	0.34916	0.55899	0.45512	0.12791
4	0.38741	0.52381	0.42899	0.14258
5	0.43838	0.47652	0.39806	0.16054
6	0.59816	0.35314	0.32324	0.16432

since $\boldsymbol{\theta}_2$ is unknown, one can adopt a pseudo-Bayesian approach by assuming some prior on $\boldsymbol{\theta}_2$ (Kiefer 1973). Then, the objective function can be shown to be a function of $\mathcal{E}(\boldsymbol{\theta}_2\boldsymbol{\theta}_2')$. One can now study the problem of finding designs for different forms and values of $\mathcal{E}(\boldsymbol{\theta}_2\boldsymbol{\theta}_2')$.

In this context, it may be mentioned that a related problem of finding optimum designs for the *discrimination* of two models in a regression problem was first considered by Atkinson and Fedorov (1975a). Later, a number of authors worked in this area and extended it to several models, (see e.g. Atkinson and Fedorov (1975b), Dette (1994), Dette and Titoff (2009), Dette et al. (2012), Fedorov and Khabarov (1986) , Tsai et al. (2004), Tommasi and López-Fidalgo (2010), Wiens (2009) etc.). It seems, the problem of finding optimum designs for the discrimination of different models in a mixture experiment is still open.

12.2 Optimum Mixture Designs for Random Coefficients Mixture Models

12.2.1 Preliminaries

In most of the investigations on regression experiment, it is assumed that the response function can be approximated by a polynomial of certain degree in the levels of the factors with fixed regression coefficients. Liski et al. (1997, 2002) were the first to consider a single-factor experiment, where the response function is approximated by a first-degree and a second-degree random coefficients regression (RCR) model. In this section, we consider the problem of determining optimum designs in a mixture experiment with RCR model defining the response function. Pal et al. (2010) and Pal and Mandal (2012b) obtained optimum designs for Scheffé's linear and quadratic mixture models with two and three components and for q-component D–W model, respectively, using A- and D- optimality criteria.

Consider the following linear and quadratic models in q mixture components

$$y = \beta_0^* + \sum_{i=1}^{q} \beta_i^* x_i + \varepsilon \tag{12.2.1}$$

and

$$y = \beta_0^* + \sum_{i=1}^{q} \beta_i^* x_i + \sum_{i=1}^{q} \sum_{j=i}^{q} \beta_{ij}^* x_i x_j + \varepsilon, \tag{12.2.2}$$

where ε is the random error with mean 0 and variance σ^2.

For fixed β_i^*s and β_{ij}^*s, optimum designs for the estimation of regression coefficients have already been discussed in Chap. 4. Let us now assume that β_0^* is random, and $\beta_i^* = \beta_i + b_i, i = 1(1)q$, where β_is are fixed and b_is are independent random, with $E(b_i) = 0$ and $\text{Var}(b_i) = \sigma_i^2$. Similarly, $\beta_{ij}^* + b_{ij}, i \leq j = 1(1)q$, with β_{ij}s fixed and b_{ij}s independent random, having $E(b_{ij}) = 0$ and $\text{Var}(b_{ij}) = \sigma_{ij}^2$. Also, we assume b_is and b_{ij}s to be independent, and independent of ε.

Then, (12.2.1) and (12.2.2) can be written as

$$y = \beta_0^* + \sum_{i=1}^{q} \beta_i x_i + \varepsilon^* \tag{12.2.3}$$

and

$$y = \beta_0^* + \sum_{i=1}^{q} \beta_i x_i + \sum_{i=1}^{q} \sum_{j=i}^{q} \beta_{ij} x_i x_j + \varepsilon^*, \tag{12.2.4}$$

where

$$\begin{aligned} \varepsilon^* &= \sum_{i=1}^{q} b_i x_i + \varepsilon, \text{ in } (12.2.3) \\ &= \sum_{i=1}^{q} b_i x_i + \sum_{i=1}^{q} \sum_{j=i}^{q} b_{ij} x_i x_j + \varepsilon, \text{ in } (12.2.4) \end{aligned}$$

Because of the constraint $\Sigma x_i = 1$, we can equivalently write (12.2.3) and (12.2.4) in the canonical form:

$$y = \sum_{i=1}^{q} \beta_i^{**} x_i + \varepsilon^* \tag{12.2.5}$$

$$y = \sum_{i=1}^{q} \sum_{j=i}^{q} \beta_{ij}^{**} x_i x_j + \varepsilon^* \tag{12.2.6}$$

where

$$\begin{aligned} \beta_{ij}^{**} &= \beta_0^* + \beta_i, i = 1(1)q \\ \beta_{ij}^{**} &= \beta_0^* + \beta_i + \beta_{ij}, j = 1(1)q. \end{aligned}$$

Then, given an $n-$point design D_n, the observational equations for the modified model can be written as

$$E(\boldsymbol{y}) = X\boldsymbol{\beta}, \tag{12.2.7}$$

where X is the coefficient matrix and $\boldsymbol{\beta}$ is the vector of fixed regression coefficients.

Further,

$$\Sigma = \text{Disp}(Y) = X\Lambda X' + \sigma^2 I_n, \tag{12.2.8}$$

with $\Lambda = \text{Diag}(\sigma_1^2, \sigma_2^2, \ldots, \sigma_q^2)$, for linear regression
$= \text{Diag}(\sigma_{11}, \sigma_{22}, \ldots, \sigma_{qq}, \sigma_{12}, \sigma_{13}, \ldots, \sigma_{q-1,q})$, for quadratic regression.

The generalized least squares estimator (GLSE) $\hat{\boldsymbol{\beta}}$ of $\boldsymbol{\beta}$ has the dispersion matrix

$$\text{Disp}(\hat{\boldsymbol{\beta}}) = (X'\Sigma^{-1}X)^{-1} \tag{12.2.9}$$

Making use of the fact that under the form (12.2.8) of Σ, ordinary least squares estimator (OLSE) and the GLSE are identical (cf. Rao 1965), one can write

$$\text{Disp}(\hat{\boldsymbol{\beta}}) = \sigma^2[\Lambda^* + (X'X)^{-1}] \tag{12.2.10}$$

where $\Lambda^* = (\sigma^2)^{-1}\Lambda$.

Pal et al. (2010) worked with the setup of a continuous design ξ for which

$$X'X = \sum_{u=1}^{n} w_u x_u x_u', \tag{12.2.11}$$

where $x_1, x_2, \ldots, x_n$ are the n support points of the design and $w_1, w_2, \ldots, w_n$ ($w_i \leq 0$, $\Sigma w_i = 1$) are the respective masses.

12.2.2 Characterization

From (12.2.10), when the variance components are all equal, the problem becomes invariant with respect to the different components, so that one can restrict to the class of symmetric designs. In particular, for the cases of two and three components, by virtue of explicit demonstration of the dominance of weighted centriod designs, as in Draper and Pukelsheim (1999), the optimum design will belong to the class of *WCD*s.

As mentioned earlier, the class of *WCD*s for q components is given by

$$C = \left\{\alpha_1\eta_1 + \alpha_2\eta_2 + \cdots + \alpha_q : 0 \leq \alpha_i \leq 1, i = 1(1)q, \sum_{i=1}^{q} \alpha_i = 1\right\},$$

where η_i is a class of design points, each having i non-zero elements of equal value and weight $\alpha_i / C(q, i)$, $i = 1(1)q$.

When σ_i^2s or σ_{ij}s are not all equal, the problem is no longer invariant with respect to the different components. However, this does not pose any threat when the choice of optimum design is made on the basis of A-optimality criterion. This can be seen by noting that

$$\text{Trace } (X'\Sigma^{-1}X)^{-1} = \sigma^2 \Big[\text{ Trace } (\Lambda^*) + \text{ Trace } (X'X)^{-1} \Big] \tag{12.2.12}$$

[(cf. Eqs. (12.2.9)–(12.2.10)]. The above breakup shows that the A-optimum design is independent of Λ^*. This is a general observation known to the optimal design theorists dealing with such models.

For other criteria, determination of the optimum design in the general setup becomes rather difficult. When using the D-optimality criterion, we restrict our investigations to the cases of two- and three-component mixtures.

For the D-optimality criterion, we have to minimize

$$\phi(\xi) = \mid \text{ Disp } (\hat{\boldsymbol{\beta}}) \mid = \sigma^{2q} \mid \Lambda^* + (X'X)^{-1} \mid. \tag{12.2.13}$$

Now, $X'X$ stands for the information matrix of the corresponding fixed effects modesl. So, for given $\boldsymbol{\Lambda^*}$, we can reduce the class of competing designs using Loewner order dominance on $X'X$.

(i) **Linear regression:** For the linear case with two components, because of the constraint $x_1 + x_2 = 1$, the response function can be reduced to a simple linear regression with one variable. Since the parameter vectors in the two representations are related through a non-singular matrix, in terms of Loewner domination, one can work with any one of the two representations. Then, using the result of Liski et al. (2002), one can restrict to the class of two-point designs with support points at $(1, 0)$ and $(0, 1)$ for the estimation of $\boldsymbol{\beta}$.

Consider an arbitrary design $\xi_0(\alpha)$ which puts mass α at $(1, 0)$ and mass $(1 - \alpha)$ at $(0, 1)$. Then, it is easy to calculate that

$$\mid \text{Disp}(\hat{\beta}) \mid = \sigma^4 \left(\frac{1}{\alpha} + \frac{\sigma_1^2}{\sigma^2} \right) \left(\frac{1}{1-\alpha} + \frac{\sigma_2^2}{\sigma^2} \right),$$

which is minimized at

$$\alpha = \alpha_0 = \frac{\sqrt{\sigma^2 + \sigma_2^2}}{\sqrt{\sigma^2 + \sigma_1^2} + \sqrt{\sigma^2 + \sigma_2^2}}. \tag{12.2.14}$$

Theorem 12.2.1 *In a first-degree two-component mixture model with random regression coefficients, having variance components σ_{11}^2 and σ_{22}^2, the D-optimal design for the estimation of fixed effects parameters is a two-point design having support points at* (1,0) *and* (0,1), *with masses α_0 and $1 - \alpha_0$, respectively, where α_0 is given by* (12.2.14).

Remark 12.2.1 Whether the regression coefficients are fixed or random, in the first-degree two-component model, the two vertices/extreme points are the only support points. Moreover, for the random regression coefficients, the mass at the vertices decreases with increasing variability of the corresponding regression coefficients.

For the case of three components, the problem becomes rather difficult when the variance components σ_i^2s are all unequal. For simplicity sake, Pal et al. (2010) assumed that

$$\sigma_1^2 = \sigma_2^2 (\neq \sigma_3^2). \tag{12.2.15}$$

In view of the optimality criterion, assumption (12.2.15) amounts to stating that the first two mixing components are 'exchangeable,' since under the transformation $X \to X^\tau = XP$, where P is a permutation matrix given by $P = (e_2, e_1, e_3)$, and e_i is the i-th unit vector, the criterion function remains invariant. This leads to the *heuristic* argument that it may be enough to search for an optimum design in the hyperplane manifested by the property of exchangeability of the first two components. It turns out that in such a plane, the linear response function involving all the three mixing components may be reduced to a linear function in the third component only. Appealing to Liski et al. (2002), one can restrict to the class of designs with x_3 taking the two values 0 and 1. This is because, from (12.2.9) and (12.2.10), dominance of $(X'\Sigma^{-1}X)^{-1}$ is equivalent to dominance of $(X'X)^{-1}$, for fixed Λ^*.

When $x_3 = 0$, the response function becomes a linear function in a single variable x_1, say, because of the constraint $x_1 + x_2 + x_3 = 1$. Then, by Liski et al. (2002), the support points must be at $x_1 = 0$ or 1. For $x_3 = 1$, clearly $x_1 = x_2 = 0$. Thus, the class of competing designs reduces to

x_1	x_2	x_3	Mass
1	0	0	αw_1
0	1	0	αw_1
1/2	1/2	0	$(1-2\alpha)w_1$
0	0	1	w_2
$1-\alpha$	0	α	$w_3/2$
0	$1-\alpha$	α	$w_3/2$

Since the problem is invariant with respect to x_1 and x_2, under the assumption (12.2.15), the masses at (1, 0, 0) and (0, 1, 0) should be equal. This is reflected in the above table.

Then,

$$| \operatorname{Disp}(\hat{\boldsymbol{\beta}}) | = (\sigma^2)^3 \left(\frac{2}{\alpha} + \frac{\sigma_1^2}{\sigma^2} \right) \left(\frac{1}{1-\alpha} + \frac{\sigma_3^2}{\sigma^2} \right). \tag{12.2.16}$$

Equation (12.2.16) is minimized at $\alpha = \alpha_0$, where

$$\alpha_0 = \frac{1}{(\sigma_1^* - 4\sigma_3^*)}\left[\left\{(3+4\sigma_3^*)^3 + 4(\sigma_1^* - 4\sigma_3^*)(1+\sigma_3^*)\right\}^{1/2} - (3+4\sigma_3^*)\right], \tag{12.2.17}$$

$$\sigma_1^* = \frac{\sigma_1^2}{\sigma^2}, \sigma_3^* = \frac{\sigma_3^2}{\sigma^2}.$$

Thus, we have the following theorem:

Theorem 12.2.2 *In a first-degree three-component mixture model with random regression coefficients having variance components* $\sigma_1^2 = \sigma_2^2, \sigma_3^2$, *the D-optimal design for the estimation of fixed effects parameters is a three-point design having support points at (1, 0, 0), (0, 1, 0) and (0, 0, 1), with mass* $\alpha/2, \alpha/2$, *and* $1-\alpha_0$, *respectively, where* α_0 *is given by* (12.2.17).

As in the two-component case, here also, the mass at the vertices decreases with increasing variability of the corresponding regression coefficients.

(ii) **Quadratic regression model**

Case of two components: For the case of two components, (12.2.5) can be equivalently written as

$$y = \delta_0 + \delta_1 x_1 + \delta_2 x_1^2 + \varepsilon^*, \quad 0 \le x_1 \le 1, \tag{12.2.18}$$

so that for a given continuous design ξ, the information matrix for the parameter vector in (12.2.6) will be a one-to-one function of that in (12.2.18).

It is known that for a single-variable quadratic response function (cf. Liski et al. 2002), given any arbitrary design ξ, there exists a three-point design ξ_0 with support points at 0, 1 and $a \in (0, 1)$, such that $(X_\delta' X_\delta) \mid_{\xi_0} - (X_\delta' X_\delta) \mid_{\xi}$ is non-negative definite, where X_δ represents the coefficient matrix in the model (12.2.18). Because of the one-to-one correspondence of the information matrices of the parameters in (12.2.6) and (12.2.18), Pal et al. (2010) restricted to the class of three-point designs with support points at $(1, 0)$, $(0, 1)$, and $(a, 1-a)$ for some $a \in (0, 1)$.

Consider an arbitrary three-point design $\xi_0(\alpha)$, which puts mass α_1 at $(1, 0)$, mass α_2 at $(0, 1)$ and mass $(1-\alpha)$ at $(a, 1-a)$, where $\alpha = \alpha_1 + \alpha_2$, $\boldsymbol{\alpha} = (\alpha_1, \alpha_2)'$. Then, the authors showed that for such a design

$$\mid \text{Disp}\,(\hat{\boldsymbol{\beta}}) \mid \ge A(\boldsymbol{\alpha}, a) + \frac{1}{\alpha}\left(\sqrt{B(\boldsymbol{\alpha}, a)} + \sqrt{C(\boldsymbol{\alpha}, a)}\right)^2, \tag{12.2.19}$$

where

$$A(\boldsymbol{\alpha}, a) = a^*(\boldsymbol{\alpha}, a)\sigma_{11}^*\sigma_{22}^*,$$

$$B(\boldsymbol{\alpha}, a) = \left[a^*(\boldsymbol{\alpha}, a) + \left(\frac{a}{1-a}\right)^2 \sigma_{11}^*\right]\sigma_{22}^* + \frac{b^*(\boldsymbol{\alpha}, a)}{\alpha},$$

Table 12.3 Optimal designs and the minimum value of the criterion function for different combinations of variance components

σ_{11}	σ_{12}	σ_{22}	a	α_1	α_2	$1-\alpha$	Min$\|$Disp $(\hat{\beta})\|$
2	0	3	0.5041	0.2986	0.2679	0.4335	1433.949
2	1	3	0.5044	0.3011	0.2699	0.4290	1469.794
2	0	5	0.5106	0.3108	0.2392	0.4500	1869.136
2	1	5	0.5104	0.3136	0.2409	0.4455	1916.820

$$C(\boldsymbol{\alpha}, a) = \left[a^*(\boldsymbol{\alpha}, a) + \left(\frac{1-a}{a}\right)^2 \sigma_{22}^*\right]\sigma_{11}^* + \frac{b^*(\boldsymbol{\alpha}, a)}{\alpha},$$

$$a^*(\boldsymbol{\alpha}, a) = \sigma_{12}^* + \frac{1}{(1-\alpha)a^2(1-a)^2},$$

$$b^*(\boldsymbol{\alpha}, a) = a^*(\boldsymbol{\alpha}, a) + \left(\frac{a}{1-a}\right)^2 \sigma_{11}^* + \left(\frac{1-a}{a}\right)^2 \sigma_{22}^*,$$

$$\sigma_{ii}^* = \sigma_{ii}/\sigma^2, \quad i = 1, 2.$$

For given $\boldsymbol{\alpha}$ and a, equality in (12.2.19) holds for

$$\alpha_1 = \frac{\sqrt{B(\boldsymbol{\alpha}, a)}}{\sqrt{B(\boldsymbol{\alpha}, a)} + \sqrt{C(\boldsymbol{\alpha}, a)}}\alpha, \quad \alpha_2 = \frac{\sqrt{C(\boldsymbol{\alpha}, a)}}{\sqrt{B(\boldsymbol{\alpha}, a)} + \sqrt{C(\boldsymbol{\alpha}, a)}}\alpha. \tag{12.2.20}$$

The optimal values of a, α_1, and α_2 are obtained so as to minimize (Table 12.3)

$$A(\boldsymbol{\alpha}, a) + \frac{1}{\alpha}\left(\sqrt{B(\boldsymbol{\alpha}, a)} + \sqrt{C(\boldsymbol{\alpha}, a)}\right)^2.$$

Case of three components: For the case of three components, analogous to the linear regression case, we assume that

$$\sigma_{11} = \sigma_{22} (\neq \sigma_{33}), \sigma_{13} = \sigma_{23} (\neq \sigma_{12}) \tag{12.2.21}$$

Assumption (12.2.21) implies that the first two mixing components are 'exchangeable,' since under the transformation $X \to X^P = XP$, where P is a permutation matrix given by $P = (\boldsymbol{e}_2, \boldsymbol{e}_1, \boldsymbol{e}_3, \boldsymbol{e}_4, \boldsymbol{e}_5, \boldsymbol{e}_6)$, and $\boldsymbol{e}_i$ is the i-th unit vector, the criterion function remains invariant. Then, arguing as in the case of three-component mixture with linear regression, we adopt an initial design with x_3 taking the three values 0, 1, and some $a \in (0, 1)$, by virtue of Liski et al. (2002). Since ln $\phi(\xi)$ is convex with respect to the $X'X$ the optimum design will be invariant with respect to the first two components. Hence, we propose the following class $\mathcal{C}$ of designs with support points as below:

x_1	x_2	x_3	Mass
1	0	0	αW_1
0	1	0	αW_1
1/2	1/2	0	$(1-2\alpha)W_1$
0	0	1	W_2
$1-a$	0	a	$W_3/2$
0	$1-a$	a	$W_3/2$

Since the problem is invariant with respect to x_1 and x_2, the masses at $(1, 0, 0)$ and $(0, 1, 0)$ should be equal, and so also the masses at $(1-a, 0, a)$ and $(0, 1-a, a)$.

Determinant of Disp $(\hat{\boldsymbol{\beta}})$ comes out to be a complicated function of a, α, W_1, W_2, and W_3, so that one cannot obtain their optimal values in closed forms. However, for given sets of values of the variance components, one can easily find the optimum design using statistical package.

12.2.2.1 An Alternative Approach

We can write model (12.2.6) as

$$\begin{aligned} y &= \theta_{11}x_1(x_1 - 1 + a) + \theta_{22}x_2(x_2 - 1 + a) + \theta_{33}x_3(x_3 - a) \\ &\quad + \theta_{12}x_1x_2 + \theta_{13}x_1x_3 + \theta_{23}x_2x_3 + \varepsilon^*, \end{aligned} \tag{12.2.22}$$

where $\boldsymbol{\beta}$ and $\boldsymbol{\theta} = (\theta_{11}, \theta_{22}, \theta_{12})'$ are related by

$$\boldsymbol{\beta} = L\boldsymbol{\theta}, \tag{12.2.23}$$

with

$$L = \begin{pmatrix} a & 0 & 0 & 0 & 0 & 0 \\ 0 & a & 0 & 0 & 0 & 0 \\ 0 & 0 & 1-a & 0 & 0 & 0 \\ -(1-a) & -(1-a) & 0 & 1 & 0 & 0 \\ -(1-a) & 0 & -a & 0 & 1 & 0 \\ 0 & -(1-a) & -a & 0 & 0 & 1 \end{pmatrix}. \tag{12.2.24}$$

Thus, there is a one-to-one relation between $\boldsymbol{\beta}$ and $\boldsymbol{\theta}$, and hence between their estimates.

Within the class $\mathcal{C}$ of designs, it can be shown that

$$\mid \text{Disp}(\hat{\boldsymbol{\beta}}) \mid = \sigma^2 a^4 (1-a)^2 C_1(W_1, a, \alpha) C_2(W_2, W_3, a), \tag{12.2.25}$$

Table 12.4 Optimum designs for different combinations of variance components (relative to the error variance)

$\sigma_{11} = \sigma_{22}$	σ_{12}	σ_{33}	$\sigma_{13} = \sigma_{23}$	a	α	w_1	w_2	w_3
0.4	2.0	0.75	0.5	0.4999	0.3338	0.4991	0.1642	0.3367
0.4	2.0	1.5	1.0	0.5029	0.3338	0.5020	0.1608	0.3372
1.0	0.5	4.0	1.0	0.5061	0.3292	0.5055	0.1512	0.3433
1.0	0.5	2.0	1.0	0.4995	0.3294	0.4934	0.1614	0.3452
1.0	0.5	0.75	0.5	0.4941	0.3295	0.4965	0.1649	0.3386
2.0	1.0	1.0	0.5	0.4871	0.3262	0.4914	0.1654	0.3432
2.0	2.0	1.0	0.5	0.4870	0.3274	0.4906	0.1656	0.3438
4.0	2.0	1.5	1.0	0.4761	0.3209	0.4838	0.1658	0.3504
10.0	1.0	1.0	1.0	0.4482	0.3062	0.4646	0.1737	0.3617
10.0	1.0	15.0	5.0	0.4764	0.3012	0.4942	0.1298	0.3760

where

$$C_1(W_1, a, \alpha) = \left(\sigma_{11} + \frac{1}{W_1 a^2 \alpha}\right)\left[\sigma_{11}\sigma_{12} + \frac{A_1}{\alpha} + \frac{A_2}{1-2\alpha}\right], \qquad (12.2.26)$$

$$C_2(W_2, W_3, a) = \left(\frac{1}{W_2(1-a)^2} + \sigma_{33}\right)\left(\frac{2}{W_3 a^2 (1-a)^2} + \sigma_{13}\right)^2, \qquad (12.2.27)$$

A_1 and A_2 being functions of W_1 and a, given by

$$A_1 = \frac{1}{a^2 W_1}\left[\sigma_{12} + 16\sigma_{11}(a - 1/2)^2 + 16/W_1\right], \quad A_2 = \frac{16}{W_1}\left[\sigma_{11} + \frac{2}{a^2 W_1}\right]. \qquad (12.2.28)$$

Given a and W_1, the optimal values of α and W_3 satisfy (Table 12.4)

$$\frac{1-2\alpha}{\alpha} = \frac{\left(\sigma_{11}\alpha + \frac{2}{W_1 a^2}\right)\left[-2\sigma_{11}\sigma_{12}\alpha^2 + (\sigma_{11}\sigma_{12} - 2A_1 + A_2)\alpha + A_1\right]}{\left(\sigma_{11}\alpha + \frac{1}{W_1 a^2}\right)\left[(\sigma_{11}\sigma_{12} - 2A_1 + A_2)\alpha + 2A_1\right]} \qquad (12.2.29)$$

$$\frac{W_3}{1-W_1} = \frac{8\sigma_{33} + \frac{6}{(1-W_1)(1-a)^2} - 16\left(4\sigma_{33} - a^2\sigma_{13}\right)\left(\sigma_{33} + \frac{1}{(1-W_1)(1-a)^2}\right)}{2\left(4\sigma_{33} - a^2\sigma_{13}\right)}. \qquad (12.2.30)$$

12.2.3 Darroch–Waller Model

Pal and Mandal (2012b) extended the concept of random regression coefficients to D–W model. An additive quadratic mixture model was introduced by Darroch and Waller (1985) for the case of $q = 3$:

$$y = \sum_{i=1}^{q} \beta_i^* x_i + \sum_{i=1}^{q} \beta_{ii}^* x_i(1 - x_i) + \varepsilon, \tag{12.2.31}$$

where ε is the random error distributed with mean 0 and variance σ^2. The details about the model and optimality considerations are considered in Chap. 6. Pal and Mandal (2012b) considered the problem of determining D-optimal design for the estimation of expected regression coefficients in the additive mixture model due to Darroch and Waller, assuming random regression coefficients.

In (12.2.31), as in Scheffé model, it is assumed that $\beta_i^* = \beta_i + b_i$, $\beta_{ii}^* = \beta_{ii} + b_{ii}$, $i = 1(1)q$, where β_is and β_{ii}s are fixed and b_is and b_{ii}s are independent random, with $E(b_i) = 0$, $E(b_{ii}) = 0$ and $\text{Var}(b_i) = \sigma_i^2$, $\text{Var}(b_{ii}) = \sigma_{ii}^2$. Also, we assume that b_is and b_{ij}s are independent of ε.

Then, we can write (12.2.31) as

$$y = \sum_{i=1}^{q} \beta_i x_i + \sum_{i=1}^{q} \beta_{ii} x_i(1 - x_i) + \varepsilon^*, \tag{12.2.32}$$

where

$$\varepsilon^* = \sum_{i=1}^{q} b_i x_i + \sum_{i=1}^{q} b_{ii} x_i(1 - x_i) + \varepsilon.$$

Given an n-point design D_n, the observational equations for the modified model can be, therefore, written as

$$E(\mathbf{y}) = X\boldsymbol{\beta}, \tag{12.2.33}$$

where $\mathbf{y}$ is the response vector, X the coefficient matrix, and $\boldsymbol{\beta}$ the vector of fixed regression coefficients.

Hence,

$$\Sigma = \text{Disp}\,(\mathbf{y}) = X\Lambda X' + \sigma_\varepsilon^2 \mathbf{I}_n, \tag{12.2.34}$$

with

$$\Lambda = \text{Diag}\left(\sigma_1^2, \sigma_2^2, \ldots, \sigma_q^2, \sigma_{11}^2, \sigma_{22}^2, \ldots, \sigma_{qq}^2\right).$$

It is assume that $X'X$ is non-singular. Since under the form (12.2.34) of Σ, OLSE and the GLSE of $\boldsymbol{\beta}$ are identical (cf. Rao 1967),

$$\text{Disp}\,(\hat{\boldsymbol{\beta}}) = \sigma_\varepsilon^2[\Lambda^* + (X'X)^{-1}] \tag{12.2.35}$$

where $\Lambda^* = (\sigma_\varepsilon^2)^{-1}\Lambda$.

12.2.3.1 *D*-optimal designs

The moment matrix for a continuous design ξ with mass w_u at the support point $\boldsymbol{x}_u$, $u = 1, 2, \ldots, n$ is given by

$$X'X = \sum_{u=1}^{n} w_u \boldsymbol{f}(\boldsymbol{x}_u)\boldsymbol{f}'(\boldsymbol{x}_u) \tag{12.2.36}$$

where $\boldsymbol{f}'(\boldsymbol{x}_u) = (x_{u1}, x_{u2}, \ldots, x_{uq}, x_{u1}(1-x_{u1}), x_{u2}(1-x_{u2}), \ldots, x_{uq}(1-x_{uq}))$, $u = 1, 2, \ldots, n$.

In order to find the *D*-optimal design for estimation of $\boldsymbol{\beta}$, it is assumed that $\sigma_i^2 = \lambda_1$ and $\sigma_{ii}^2 = \lambda_2$, for all i. The assumption implies that the problem is invariant with respect to the components x_is so that the optimum design must be an invariant design.

For the second-degree Scheffé model, Draper et al. (2000) established that given any arbitrary symmetric design ξ, there exists a WCD which dominates ξ in the Loewner order sense. Since model (1.2) can be written as a special case of Scheffé's quadratic model (1.1) with $\theta_i = \beta_i$ and $\theta_{ij} = \beta_{jj}$, for all $i, j = 1, 2, \ldots, q, i < j$, Loewner order of the information matrices for the Scheffé model also holds for those of the model (12.2.32). Hence, using the result of Draper et al. (2000), in order to find the *D*-optimal design, we may restrict to the class of WCDs.

Remark 12.2.2 From the above argument, it follows that the search for optimal design may be restricted to the class of WCDs for all optimality criteria, which are functions of the information matrix, convex with respect to it, and are invariant with respect to its components.

Remark 12.2.3 From (2.5), it is clear that the *A*-optimal design will be same as that for the corresponding fixed effects model.

For a q-component mixture model, writing α_j to be the mass at each barycenter of depth $(j-1)$, it can be easily checked that the moment matrix of a WCD is obtained in the form

$$X'X = \begin{bmatrix} a_1 I_q + a_2 J_q & b_1 I_q + b_2 J_q \\ b_1 I_q + b_2 J_q & c_1 I_q + c_2 J_q \end{bmatrix},$$

where a_is, b_is and c_is are linear functions of the weights α_js.

We, therefore, obtain $X'X^{-1}$ in the form

$$(X'X)^{-1} = \begin{bmatrix} e_1 I_q + e_2 J_q & d_1 I_q + d_2 J_q \\ d_1 I_q + d_2 J_q & g_1 I_q + g_2 J_q \end{bmatrix},$$

where e_is, f_is and d_is are non-linear functions of α_js.

Pal and Mandal (2012b) numerically computed α_js for several combinations of (λ_1, λ_2) and for $q = 3, 4, \ldots, 10$. The following observations have been made from the study:

1. For all $3 \leq q \leq 10$, barycenters of depth at most 2 form the support points of the D-optimal design. Among these, the barycenters of depth 0 are necessarily the support points of the optimal design.
2. For $q = 3, 4$, the optimal design assigns positive weights to barycenters of depths 0 and 1 only. Thus, for $q = 3$, the D-optimal design is saturated.
3. For $5 \leq q \leq 7$, the optimal design can assign positive weights to barycenters of depths 0, 1, and 2 only. However, for given λ_1 as λ_2 increases, the weight at barycentres of depth 2 decreases and tends toward zero while for given λ_2, as λ_1 increases, the weight at barycentres of depth 1 tends toward zero.
4. For $8 \leq q \leq 10$, the optimal design assigns positive weights to barycenters of depths 0 and 2 only.

Table 12.5 gives the D-optimal designs for $q = 3, 4, \ldots, 10$ for some combinations of (λ_1, λ_2).

From Table 12.5, we note that for $4 \leq q \leq 10$, the D-optimal designs are not saturated. It would, therefore, be interesting to compare the performance of optimum saturated designs as against the optimum designs in the entire class of competing designs. Pal and Mandal (2012b) considered a simple subclass of saturated designs D_s based on barycentres of depths 0 and $(q-2)$, for a q-component mixture experiment.

In Table 12.6, the efficiency factor of the D-optimum design in D_s with respect to the optimum design given in Table 12.5, for $4 \leq q \leq 6$.

Remark 12.2.4 The above table shows that the optimum design in the entire class of competing designs is highly efficient in comparison with the optimum design in the saturated class D_s.

One can also consider another subclass of saturated designs by making use of a subset of barycenters of depth 1, and check the relative performance of the D-optimal design in it as against the D-optimal design in the whole class of competing designs.

Let us consider $q = 4$. In this case, a reasonable choice of a subset of 4 barycentres of depth 1 can be made by deleting two points which have either no or only one component in common. Let D_{s1} and D_{s2} denote, respectively, the two subclasses of designs thus obtained. In the former situation, the information matrix is singular. In the second case, the efficiency of the best design in D_{s2} relative to the optimum design in the entire class is found to be sufficiently small. The efficiency for some combinations of (λ_1, λ_2) is shown in Table 12.7.

Remark 12.2.5 Comparing the efficiency factors obtained in Table 12.6 and in Table 12.7 for $q = 4$, it is evident that the D-optimal design in D_{s2} performs better than that in D_s.

Table 12.5 *D*-optimal designs for some combinations of (λ_1, λ_2)

q	λ_1	λ_2	$C(q,1)\alpha_1$	$C(q,2)\alpha_2$	$C(q,3)\alpha_3$
3	0	0	0.5000	0.5000	0
	1	1	0.4848	0.5152	0
	1	5	0.4994	0.5006	0
	1	10	05149	0.4851	0
	5	1	0.4449	0.5551	0
	5	5	0.4535	0.5465	0
	5	10	0.4633	0.5367	0
4	0	0	0.5000	0.5000	0
	1	1	0.4890	0.5110	0
	1	5	0.5254	0.4746	0
	1	10	0.5590	0.4410	0
	5	1	0.4605	0.5395	0
	5	5	0.4749	0.5251	0
	5	10	0.4911	0.5089	0
5	0	0	0.4946	0.4077	0.0977
	1	1	0.4882	0.4265	0.0853
	1	5	0.4994	0.4368	0.0638
	1	10	0.5117	0.4440	0.0443
	5	1	0.4524	0.3343	0.2133
	5	5	0.4613	0.3524	0.1863
	5	10	0.4712	0.3688	0.1600
	5	200	0.6302	0.3698	0
	50	10	0.3647	0	0.6353
6	0	0	0.4959	0.2753	0.2288
	1	1	0.4869	0.2350	0.2781
	1	5	0.4967	0.2554	0.2479
	1	10	05076	0.2731	0.2193
	5	1	0.4538	0.0837	0.4625
	5	5	0.4616	0.1094	0.4290
	5	10	0.4705	0.1343	0.3952
	50	10	0.3764	0	0.6236
7	0	0	0.4977	0.0877	0.4146
	1	1	0.4903	0.0397	0.4700
	1	5	0.4987	0.0640	0.4373
	1	10	0.5082	0.0868	0.4050
	5	1	0.4593	0	0.5407
	5	5	0.4666	0	0.5334
	5	10	0.4748	0	0.5252
8	0	0	0.5000	0	0.5000
	1	1	0.4923	0	0.5077
	1	5	0.5004	0	0.4996
	1	10	0.5096	0	0.4904
	5	1	0.4634	0	0.5366
	5	5	0.4703	0	0.5297
	5	10	0.4778	0	0.5222

(continued)

Table 12.5 (continued)

q	λ_1	λ_2	$C(q,1)\alpha_1$	$C(q,2)\alpha_2$	$C(q,3)\alpha_3$
9	0	0	0.5000	0	0.5000
	1	1	0.4932	0	0.5068
	1	5	0.5005	0	0.4995
	1	10	0.5089	0	0.4911
	5	1	0.4667	0	0.5333
	5	5	0.4728	0	0.5272
	5	10	0.4801	0	0.5199
10	0	0	0.5000	0	0.5000
	1	1	0.4938	0	0.5062
	1	5	0.5005	0	0.4995
	1	10	0.5083	0	0.4917
	5	1	0.4694	0	0.5406
	5	5	0.4751	0	0.5249
	5	10	0.4819	0	0.5181

Table 12.6 D-optimum design in D_s for some combinations of (λ_1, λ_2) and its efficiency for $4 \le q \le 6$

q	λ_1	λ_2	$C(q,1)\alpha_1$	$C(q,q-1)\alpha_{q-1}$	Efficiency
4	0	0	0.5000	0.5000	7.40×10^{-2}
	1	1	0.4854	0.5146	3.95×10^{-1}
	1	5	0.4983	0.5017	3.87×10^{-1}
	1	10	0.5111	0.4889	3.90×10^{-1}
	5	1	0.4392	0.5608	4.80×10^{-2}
	5	5	0.4474	0.5526	4.68×10^{-1}
	5	10	0.4564	0.5436	4.60×10^{-2}
5	0	0	0.5000	0.5000	3.63×10^{-3}
	1	1	0.4798	0.5202	4.54×10^{-3}
	1	5	0.4834	0.5166	5.33×10^{-3}
	1	10	0.4877	0.5123	6.41×10^{-3}
	5	1	0.4229	0.5771	7.65×10^{-3}
	5	5	0.4262	0.5738	8.69×10^{-3}
	5	10	0.4303	0.5697	1.01×10^{-2}
	5	200	0.5304	0.4696	1.12×10^{-1}
	50	10	0.2843	0.7157	3.91×10^{-2}
6	0	0	0.5000	0.5000	7.99×10^{-5}
	1	1	0.4822	0.5178	1.03×10^{-4}
	1	5	0.4843	0.5157	1.26×10^{-4}
	1	10	0.4870	0.5130	1.58×10^{-4}
	5	1	0.4295	0.5705	1.80×10^{-4}
	5	5	0.4316	0.5684	2.20×10^{-4}
	5	10	0.4341	0.5659	2.75×10^{-4}
	50	10	0.2785	0.7215	1.40×10^{-3}

Table 12.7 D-optimal design in D_{s2} and its efficiency for some combinations of (λ_1, λ_2) when $q = 4$

λ_1	λ_2	$C(q, 1)\alpha_1$	$C(q, 1)\alpha_2$	Efficiency
0	0	0.5928	0.4072	1.54×10^{-1}
1	1	0.5742	0.4258	1.75×10^{-1}
1	5	0.5784	0.4216	1.91×10^{-1}
1	10	0.5834	0.4166	2.13×10^{-1}
5	1	0.5255	0.4745	2.26×10^{-1}
5	5	0.5287	0.4713	2.39×10^{-1}
5	10	0.5325	0.4675	2.55×10^{-1}

12.3 Optimum Designs for Optimum Mixture in Some Variants of Scheffé's Quadratic Mixture Model

12.3.1 Preliminaries

The existing literature on mixture experiments mostly assumes a single response, which is dependent only on the composition of the mixture. However, there are many practical situations where the experimenter is interested in more than one characteristic feature of the output. For example, in pharmaceutical or biomedical research, though the efficacy of a drug or remedy is of primary concern, one cannot ignore the serious side-effects. In consumer products, like food and beverages, besides taste, different other aspects like color, texture, and the undesirable effects of by-products have to be taken into account. Though multi-response models have been studied in the context of response surface, (cf. Roy et al. 1971) very few such studies have been made in mixture experiments. Also, in both single and multiple response models, the response may depend not only on the mixing proportions, but also on the amount of the mixture used. An example is the effect of a fertilizer on the yield of a crop, which depends on the composition of the fertilizer as well as on the amount of fertilizer applied. To date, there are very few studies on mixture–amount models.

Optimum designs for estimation of the parameters of mixture and mixture–amount models with single response have been reviewed in Chap. 4. The present chapter focuses on the optimum designs for estimation of optimum mixture composition/amount in the single-response mixture–amount model proposed by Pal and Mandal (2012a) and also in the multi-response mixture and mixture–amount models due to Mandal and Pal (2013) and Pal and Mandal (2013), respectively. These models are basically variants of Scheffé's quadratic mixture model.

12.3.2 Optimality in Mixture–Amount Model of Pal and Mandal

Pal and Mandal (2012a) defined the response function $\eta_{x,A}$ as

$$\eta_{x,A} = \alpha_{00}^* + \alpha_{01}^* A + \alpha_{02}^* A^2 + A \sum_{i=1}^{q} \alpha_i^* x_i + \sum_{i=1}^{q} \alpha_{ii}^* x_i^2 + \sum_{i<j=1}^{q} \alpha_{ij}^* x_i x_j, \quad (12.3.1)$$

where $A \in [A_L, A_U]$, $A_L > 0$, denotes the amount and $x_1, x_2, \ldots, x_q$ are the proportions of the q components in the mixture. The assumption $A_L > 0$ ensures that some amount of the mixture is to be used in the experiment.

Imposing suitable transformation on the amount, and using the restriction $\Sigma_{i=1}^{q} x_i = 1$, one can easily rewrite (12.3.1) as

$$\eta_{x,A^*} = \beta_{00} A^{*2} + A^* \sum_i \beta_{0i} x_i + \sum_i \beta_{ii} x_i^2 + \sum_{i<j} \beta_{ij} x_i x_j = x^{*\prime} B^* x^*, \quad (12.3.2)$$

where $\boldsymbol{x}^* = (x_1, x_2, \ldots, x_q, A^*)'$ and B^* is a symmetric matrix based on $\boldsymbol{\beta}_{ij}$s, and the experimental region is given by

$$\Xi = \left\{ (x_1, x_2, \ldots, x_q, A^*) : x_i \geq 0, 1 \leq i \leq q, \sum_{i=1}^{q} x_i = 1, A^* \in [-1, 1] \right\}. \quad (12.3.3)$$

Mandal and Pal (2012) worked with the model (12.3.2). They assumed that B^* is negative definite, so that the response function is concave in x^*, and that the optimum point at which the response is maximized is an interior point of Ξ. The optimum point, subject to the constraint $c'x^* = 1$, where $\boldsymbol{c} = (1, 1, \ldots, 1, 0)'$, is obtained as $\boldsymbol{\gamma}^* = (\gamma_1^*, \gamma_2^*, ..., \gamma_q^*, A^*)' = \mu^{-1} B^{*-1} \boldsymbol{c}$, where $\mu = \boldsymbol{c}' B^{*-1} \boldsymbol{c}$. Clearly, $\boldsymbol{\gamma}^*$ is a non-linear function of the parameters of (12.3.2). As such, for any continuous design ξ with information matrix $M(\xi)$, the large sample dispersion of the estimate $\hat{\boldsymbol{\gamma}}^*$, given by $T(\boldsymbol{\gamma}^*) M^{-1}(\xi) T(\boldsymbol{\gamma}^*)'$ involves β_{ij}s as nuisance parameters through $T(\boldsymbol{\gamma}^*)$, which is the matrix of partial derivatives of $\boldsymbol{\gamma}^*$ with respect to the model parameters β_{ij}s. This was also observed in the problems discussed in Chap. 7. Further, the dispersion matrix is singular as rank $[T(\boldsymbol{\gamma}^*)] < q + 1$. For ready reference, the matrix $[T(\boldsymbol{\gamma}^*)]$ is shown below, except for a constant multiplier d^* which is independent of the design and hence, may be ignored.

$$\begin{bmatrix}
0 & -(q-1)A^*/2 & \ldots & A^*/2 & -(q-1)\gamma_1^* & \ldots & \gamma_q^* & \frac{1}{2}\gamma_1^* - \frac{q-1}{2}\gamma_2^* & \ldots & \frac{1}{2}(\gamma_{q-1}^* + \gamma_q^*) \\
0 & A^*/2 & \ldots & A^*/2 & \gamma_1^* & \ldots & \gamma_q^* & \frac{1}{2}\gamma_2^* - \frac{q-1}{2}\gamma_1^* & \ldots & \frac{1}{2}(\gamma_{q-1}^* + \gamma_q^*) \\
\ldots & \ldots & \ldots & \ldots & \ldots & \ldots & \ldots & \ldots & \ldots & \ldots \\
0 & A^*/2 & \ldots & A^*/2 & \gamma_1^* & \ldots & \gamma_q^* & \frac{1}{2}(\gamma_1^* + \gamma_2^*) & \ldots & \frac{1}{2}(\gamma_{q-1}^* - \frac{q-1}{2}\gamma_q^*) \\
0 & A^*/2 & \ldots & A^*/2 & \gamma_1^* & \ldots & \gamma_q^* & \frac{1}{2}\gamma_2^* - \frac{q-1}{2}\gamma_1^* & \ldots & \frac{1}{2}(\gamma_{q-1}^* + \gamma_q^*) \\
0 & A^*/2 & \ldots & A^*/2 & \gamma_1^* & \ldots & \gamma_q^* & \frac{1}{2}\gamma_2^* - \frac{q-1}{2}\gamma_1^* & \ldots & \frac{1}{2}(\gamma_{q-1}^* + \gamma_q^*) \\
-qA^* & -q\gamma_1^*/2 & \ldots & -q\gamma_q^*/2 & 0 & \ldots & 0 & 0 & \ldots & 0
\end{bmatrix}$$

Hence, the trace criterion is appropriate for comparing competing designs, and, making use of the pseudo-Bayesian approach of Pal and Mandal (2006) to deal with the nuisance parameters, the criterion function is given by

$$\phi(\xi) = \mathcal{E}[\text{Disp}(\hat{\boldsymbol{\gamma}}^*)] = tr[M^{-1}(\xi)\mathcal{E}(T'(\boldsymbol{\gamma}^*)T(\boldsymbol{\gamma}^*))],$$

where $\mathcal{E}$ denotes the expectation with respect to the prior distribution of $\boldsymbol{\gamma}^*$. It may be noted that $\mathcal{E}(T'(\boldsymbol{\gamma}^*)T(\boldsymbol{\gamma}^*))$ involves only the second-order moments (pure and mixed) of γ_i^*s and A^*. Assuming that there is no knowledge about the relative importance of the mixing components, and their interaction with the amount, Mandal and Pal (2012) assumed that

$$\mathcal{E}(\gamma_i^{*2}) = v, \quad i = 1, \ldots, q; \mathcal{E}(\gamma_i^* \gamma_j^*) = w, \quad i \neq j = 1, \ldots, q; w \leq v > 0,$$
$$\mathcal{E}(A^{*2}) = a, \mathcal{E}(A^* \gamma_i^*) = b, \quad i = 1, \ldots, q.$$

Clearly, $1/q^2 < v < 1/q, qv + q(q-1)w = 1$.

To obtain the trace-optimal design, we note that because of invariance in the prior moments of $\gamma_i^*, i = 1, \ldots, q, \phi(\xi)$ is invariant with respect to the mixing proportions. Hence, the optimum design will also be invariant with respect to x_is. Further, for x_is given, since the model (12.3.2) is quadratic in A^*, the optimum design is likely to admit three distinct values of A^*, two at the two extremes and one in between, with positive masses. Hence, one can initially confine the search for optimal design within the subclass $\mathcal{D}_q$ of designs having the support points and masses as given in Table 12.8, where w_{-1}, w_0, and w_1 denote the masses attached to $A^* = -1, a_0, 1$, respectively, $a_0 \in (-1, 1)$, while the sixth column gives the masses for different $(x_1, x_2, \ldots, x_q)$ combinations when A^* is given. Then, $0 \leq p_i, p_i', p_i'' \leq 1, i = 1, 2; C(q, 1)p_1 + C(q, 2)p_2 = 1, C(q, 1)p_1' + C(q, 2)p_2' = 1, C(q, 1)p_1'' + C(q, 2)p_2'' = 1, w_j > 0, j = -1, 0, 1, w_{-1} + w_0 + w_1 = 1$.

Since for quadratic regression in $[-1, 1]$, the optimum support points of D-, A-, and E-optimality criteria are at -1, 0, and 1, with equal masses at the extreme points, to start with, Mandal and Pal (2012) took $a_0 = 0$, and $w_{-1} = w_1, p_i = p_i'', i = 1, 2$. Let $\mathcal{D}_q^0 (\subset \mathcal{D}_q)$ define the corresponding subclass of designs.

Then, the criterion function comes out to be

$$\begin{aligned}
\phi(\xi) = {} & q^2 \frac{a}{2w_1(1 - 2w_1k_1)} + q \frac{a_1(k_2 + k_3) + b_1(k_2 + qk_3)}{2w_1} \\
& + q \frac{2w_1}{1 - 2w_1k_1} \left[t \left\{ t(a + qb_3) + (q-1)s(d_3 + \frac{q-2}{2}d_4) \right\} \right. \\
& \left. + \frac{q-1}{8} s^2 \left\{ d_s + 2(q-2)d_6 + \frac{(q-2)(q-3)}{2} d_7 \right\} \right] + \frac{a_3 + b_3 - (q-1)d_3}{r} \\
& + \frac{q(q-1)d_5}{8r_2} + \frac{q(q-1)}{8r} \left[2 \left\{ d_5 + (q-2)d_6 \right\} \right],
\end{aligned}$$

Table 12.8 Support points of designs in the subclass $\mathcal{D}_q$

x_1	x_2	...	x_{q-1}	x_q	Mass	A^*	Mass
1	0	...	0	0	p_1	-1	w_{-1}
0	1	...	0	0	p_1		
...	...	...	...	...	...		
0	0	...	0	1	p_1		
1/2	1/2	...	0	0	p_2		
1/2	0	...	0	0	p_2		
...	...	...	...	...	...		
0	0	...	1/2	1/2	p_2	a_0	w_0
1	0	...	0	0	p'_1		
0	1	...	0	0	p'_1		
...	...	...	...	...	...		
0	0	...	0	1	p'_1		
1/2	1/2	...	0	0	p'_2		
1/2	0	...	0	0	p'_2		
...	...	...	...	...	...		
0	0	...	1/2	1/2	p'_2	1	w_1
1	0	...	0	0	p''_1		
0	1	...	0	0	p''_1		
...	...	...	...	...	...		
0	0	...	0	1	p''_1		
1/2	1/2	...	0	0	p''_2		
1/2	0	...	0	0	p''_2		
...	...	...	...	...	...		
0	0	...	1/2	1/2	p''_2		

where $a_1, a_3, b_1, b_3, d_3, d_4, d_5, d_6, d_7, k_1, k_2, k_3, t, s, r, r_2$ are non-linear functions of the prior moments and the masses p_1, p'_1, and w_1. The optimal values of the masses are obtained by minimizing $\phi(\xi)$. In order to check for optimality or otherwise of the design thus obtained within the entire class, Kiefer's equivalence theorem is used, which, for trace-optimality criterion, reduces to Theorem 7.4.2. As the algebraic derivations are rather involved, Mandal and Pal (2012) verified the optimality condition by numerical computation, using enumerable points in the experimental region. We give below the optimum designs for q-component model, $2 \leq q \leq 4$, which have been numerically examined by Mandal and Pal (2012) to satisfy the optimality condition for some combinations of the *apriori* moments (a, v, w). The designs for $q \geq 5$ can be similarly obtained and their optimality or otherwise checked numerically using Theorem 7.4.2.

Remark 12.3.1 The above table shows that for $q = 2, 3, 4$, the optimum design assigns masses at the two extremes and the midpoint of the domain of amount A^*,

Table 12.9 Trace-optimal designs in q-component mixture–amount model for some combinations of the apriori moments (v, w, a) when $2 \leq q \leq 4$

q	a	v	p_1	p_1	w_1	w_0	Min. Trace
2	0.2	0.30	0.31985	0.30013	0.32389	0.35222	16.5421
	0.2	0.40	0.28899	0.27169	0.33789	0.32422	30.7934
	0.6	0.30	0.33841	0.31042	0.29970	0.40060	25.3021
	0.6	0.40	0.30222	0.27612	0.30860	0.38280	40.8687
3	0.2	0.20	0.12499	0.10767	0.36118	0.27763	193.945
	0.2	0.30	0.12820	0.11601	0.37395	0.25209	343.270
	0.6	0.20	0.13472	0.10401	0.33976	0.32048	225.357
	0.6	0.30	0.13416	0.11395	0.34788	0.30424	374.872
4	0.2	0.10	0.06659	0.04836	0.37601	0.24798	553.197
	0.2	0.20	0.07417	0.06460	0.39078	0.21844	1378.92
	0.6	0.10	0.07463	0.04131	0.36215	0.27570	636.485
	0.6	0.20	0.07801	0.05991	0.37113	0.25774	1461.16

which is taken to be finite, and at the support points of the $(q, 2)$ simplex design. Similar result is expected to be obtained for higher values of q (Table 12.9).

12.3.3 Optimality in Multi-response Mixture Model

Consider a mixture experiment with q-components in proportions $x_1, x_2, \ldots, x_q$, where $x_i \geq 0, i = 1(1)q, \Sigma_i x_i = 1$. Suppose the response is a p-dimensional vector $\boldsymbol{y}' = (y^{(1)}, y^{(2)}, \ldots, y^{(p)})$, where $y^{(g)}$ denotes the gth characteristic of the output, $g = 1, 2, \ldots, p$. Let, for each g, the mean response $E(y^{(g)} \mid \boldsymbol{x}) = \eta_{\boldsymbol{x}}^{(g)}$ be defined by Scheffé's quadratic mixture model:

$$\eta_{\boldsymbol{x}}^{(g)} = \beta_0^{(g)} + \sum_i \beta_i^{(g)} x_i + \sum_{i<j} \beta_{ij}^{(g)} x_i x_j, \quad g = 1, 2, \ldots, p. \tag{12.3.4}$$

Thus, as in the single-response case, using the constraint $\Sigma_i x_i = 1$, one can write for each $g = 1, 2, \ldots, p$,

$$\eta_{\boldsymbol{x}}^{(g)} = \sum_i \beta_{ii}^{(g)} x_i^2 + \sum_{i<j} \beta_{ij}^{(g)} x_i x_j = \boldsymbol{f}'(\boldsymbol{x})\boldsymbol{\beta}^{(g)} = \boldsymbol{x}'\boldsymbol{\beta}^{(g)}\boldsymbol{x}, \tag{12.3.5}$$

where $\boldsymbol{f}(\boldsymbol{x}) = (x_1^2, x_2^2, \ldots, x_q^2, x_1x_2, x_1x_3, \ldots, x_{q-1}x_q)'$, $\boldsymbol{\beta}^{(g)} = (\beta_{11}^{(g)}, \beta_{22}^{(g)}, \ldots, \beta_{qq}^{(g)}, \beta_{12}^{(g)}, \ldots, \beta_{q-1q}^{(g)})'$, and $B^{(g)} = ((b_{ij}^{(g)}))$ is a symmetric matrix with $b_{ii}^{(g)} = \beta_{ii}^{(g)}$, for all i and $b_{ij}^{(g)} = \frac{1}{2}\beta_{ij}^{(g)}$, for all $i < j$. Mandal and Pal (2013) worked with the model $\eta_{\boldsymbol{x}}^{(g)} = \boldsymbol{x}'B^{(g)}\boldsymbol{x}$, $g = 1, 2, \ldots, p$. They assumed $\eta_{\boldsymbol{x}}^{(g)}$ for each $g = 1, 2, \ldots, p$

to be concave with a finite maximum in the interior of the experimental region $\mathcal{X} = \{(x_1, x_2, \ldots, x_q) : x_i \geq 0, 0 \leq q, \sum_{i=1}^{q} x_i = 1\}$. Thus, the optimum mixture composition for the g-th response is $\boldsymbol{\gamma}^{(g)} = \delta^{(g)-1}\mathbf{1}_q$, where $\delta^{(g)} = \mathbf{1}_q' B^{(g)-1}\mathbf{1}_q$, $g = 1, 2, \ldots, p$. They studied the problem of estimating the non-linear functions $\boldsymbol{\gamma}^{(g)}$s of the unknown model parameters as accurately as possible by a proper choice of a continuous design in $\mathcal{X}$. One can write (12.3.5) as $E(\boldsymbol{y}) = \boldsymbol{\beta} \otimes \boldsymbol{f}(\boldsymbol{x})$, where $\boldsymbol{\beta} = (\boldsymbol{\beta}^{(1)\prime}, \boldsymbol{\beta}^{(2)\prime}, \ldots, \boldsymbol{\beta}^{(p)\prime})'$. Then, if $\Sigma = ((\sigma_{gh}))$ denotes the dispersion matrix of $\boldsymbol{y}$, for any arbitrary continuous design ξ in $\mathcal{X}$, the information matrix for estimating $\boldsymbol{\beta}$ is given by $I(\xi, \boldsymbol{\beta}) = \Sigma^{-1} \otimes M(\xi)$, where $M(\xi) = \int_{\mathcal{X}} \boldsymbol{f}(\boldsymbol{x})\boldsymbol{f}'(\boldsymbol{x})d\xi(\boldsymbol{x})$.

Let $\boldsymbol{\gamma} = (\boldsymbol{\gamma}^{(1)\prime}, \boldsymbol{\gamma}^{(2)\prime}, \ldots, \boldsymbol{\gamma}^{(p)\prime})'$. Then, under suitable regularity assumptions on error distribution, the standard ∂-method gives an adequate approximation of the dispersion matrix of $\hat{\boldsymbol{\gamma}} = (\hat{\boldsymbol{\gamma}}^{(1)\prime}, \hat{\boldsymbol{\gamma}}^{(2)\prime}, \ldots, \boldsymbol{\gamma}^{(p)\prime})'$ as

$$\begin{aligned}\text{Disp}(\hat{\boldsymbol{\gamma}}) = \text{Diag}(A^{(g)}(\boldsymbol{\gamma}), g = 1, 2, \ldots, p)(\Sigma \otimes M^{-1}(\xi))\text{Diag}(A^{(g)}(\boldsymbol{\gamma}),\\ g = 1, 2, \ldots, p)',\end{aligned}$$

where, for $g = 1, 2, \ldots, p$, $A^{(g)}(\boldsymbol{\gamma})$ is the matrix of partial derivatives of $\boldsymbol{\gamma}^{(g)}$ with respect to the components of $\boldsymbol{\beta}^{(g)}$ and it comes out to be a constant '$d^{(g)}$' times a matrix whose elements are linear in the components of $\boldsymbol{\gamma}^{(g)}$ (cf. Pal and Mandal 2006). The constant multiplier '$d^{(g)}$' is independent of the design ξ. In Chap. 7, these results have already been discussed.

Mandal and Pal (2013) assumed $M(\xi)$ to be negative definite. However, since $A^{(g)}(\boldsymbol{\gamma})'\mathbf{1}_q = 0$, Disp($\hat{\boldsymbol{\gamma}}$) is singular. Hence, for comparing different designs, they appropriately considered the trace of Disp($\hat{\boldsymbol{\gamma}}$). But, Disp($\hat{\boldsymbol{\gamma}}$) depends on the unknown model parameters through $A^{(g)}(\boldsymbol{\gamma})$, $g = 1, 2, \ldots, p$. From the structure of $A^{(g)}(\boldsymbol{\gamma})$, it is evident that a linear optimality criterion will be quadratic in the $\boldsymbol{\gamma}^{(g)}$-components. Therefore, to remove the nuisance parameters, assuming a prior on the first two moments of the $\boldsymbol{\gamma}^{(g)}$-components is adequate. This is precisely what was done in Pal and Mandal (2006) in the single-response model. Arguing as in Pal and Mandal (2006), it is assumed that

$$\mathcal{E}\left(\boldsymbol{\gamma}^{(g)}\boldsymbol{\gamma}^{(g)\prime}\right) = \left(v^{(g)} - w^{(g)}\right) I_q + w^{(g)}\mathbf{1}_q\mathbf{1}_q',$$

where $v^{(g)}, w^{(g)} > 0$, $w^{(g)} < v^{(g)}$, $v^{(g)} + (q-1)w^{(g)} = \frac{1}{q}$, and $v^{(g)} > \frac{1}{q^2}$.

Then, the criterion function is defined as $\phi(\xi) = \mathcal{E}[\text{Trace}(\text{Disp}(\hat{\gamma}))]$, which, after some algebraic manipulation, reduces to

$$\phi(\xi) = \text{Trace}\left[M^{-1}(\xi)\sum_{g}\sigma_{gg}^{*}\mathcal{E}\left\{A^{(g)}(\boldsymbol{\gamma})'A^{(g)}(\boldsymbol{\gamma})\right\}\right],$$

where, σ_{gg}^{*} is the product of σ_{gg} and the expected value of $d^{(g)}$.

Under the assumed prior moments, $\phi(\xi)$ is easily seen to be invariant with respect to the components of the mixture. Moreover, Trace(Disp($\hat{\boldsymbol{\gamma}}$)) is a convex function of $M(\xi)$. Hence, one can restrict the search for trace-optimal design within the subclass of invariant designs. Now, Draper and Pukelsheim (1999) and Draper et al. (2000) proved that given a symmetric design ξ, there exists a WCD η which dominates ξ in the sense of PLO. By virtue of the result and the fact that an invariant design is necessarily symmetric, the search can be further confined to the class of WCDs. Pal and Mandal (2007) showed that for $q \le 4$, the trace-optimal design in the single-response case is a $(q, 2)$-simplex lattice design. In view of that, Mandal and Pal (2013) first restricted their search within the subclass $\mathcal{D}_1$ of $(q, 2)$-simplex Lattice designs, for $q \le 4$, and then examined its optimality or otherwise within the entire class of competing designs. Using the equivalence theorem in the single-response case as stated in Pal and Mandal (2007), one can easily obtain the equivalence theorem in the multi-response case, which can be used for verification of optimality of a design:

Theorem 12.3.1 (Equivalence Theorem) *A necessary and sufficient condition for a design ξ^* to be trace-optimal in a p-variate regression model is that*

$$\boldsymbol{f}(\boldsymbol{x})'M^{-1}(\xi^*)\left[\sum_{g=1}^{p}\sigma^*_{gg}\mathcal{E}(A^{(g)\prime}A^{(g)})\right]M^{-1}(\xi^*)\boldsymbol{f}(\boldsymbol{x}) \le \phi(\xi^*) \qquad (12.3.6)$$

for all $\boldsymbol{x} \in \mathcal{X}$.
Equality in (12.3.6) *holds at all the support points of ξ^*.*

The support points of a $(q, 2)$-simplex lattice design are the vertex points: $(1, 0, 0, \ldots, 0)$ and its permutations, and the midpoints of the edges: $(1/2, 1/2, 0, \ldots, 0)$ and its permutations. Since the optimal design is invariant, we attach mass α/q to each of the vertex points and mass $(1-\alpha)/C(q, 2)$ to each of the midpoints of the edges, $0 < \alpha < 1$.

Writing the model (12.3.5) as

$$\zeta_x^{(g)} = \sum_i \theta_{ii}^{(g)} x_i(x_i - 1/2) + \sum_{i<j} \theta_{ij}^{(g)} x_i x_j, g = 1, 2, \ldots, p,$$

where

$$\boldsymbol{\theta}^{(g)} = P\boldsymbol{\beta}^{(g)}, P = \begin{bmatrix} 2I_q & 0 \\ R & I_{C(q,2)} \end{bmatrix},$$

and R is a $C(q, 2) \times q$ circulant matrix, it is easy to show that

$$\phi(\xi) = q^3(q-1)\left[\frac{t_1}{\alpha} + \frac{t_2}{1-\alpha}\right] \ge q^3(q-1)\left(\sqrt{t_1} + \sqrt{t_2}\right)^2,$$

where

$$t_1 = \sum_{g=1}^{p} \sigma_{gg}^* \left\{ (q+8)v^{(g)} + (q-1)(q-8)w^{(g)} \right\},$$

$$t_2 = 8(q-1) \sum_{g=1}^{p} \sigma_{gg}^* \left\{ (q-1)v^{(g)} - w^{(g)} \right\}.$$

Hence, $\phi(\xi)$ attains the lower bound at $\alpha = \alpha_0 = \frac{\sqrt{t_1}}{\sqrt{t_1}+\sqrt{t_2}}$.

To establish the optimality of the above designs in the entire class, it may be noted that the left-hand side of (12.3.6) is a convex function of $\boldsymbol{x}$. Hence, the maximum must occur at a boundary point of the simplex $\mathcal{X}$. It is, therefore, sufficient to verify the condition (12.3.6) only at the boundary points. However, as algebraic verification is rather involved, Mandal and Pal (2013) checked the condition numerically for $q = 3, 4, \ldots, 8$ at enumerable points on the boundary and found it to be satisfied. The following table gives the optimum designs for some combinations of $(\sigma_{gg}^*, v^{(g)}, w^{(g)})$ (cf. Mandal and Pal 2013) (Table 12.10).

12.3.4 Optimality in Multi-response Mixture–Amount Model

Pal and Mandal (2013) augmented the multi-response mixture model of Sect. 12.2 by the inclusion of amount of mixture. Let A denote the amount of mixture used. Then, for each $g = 1, 2, \ldots, p$, the mean response is assumed to be quadratic in $x^* = (A, x_1, x_2, \ldots, x_q)$:

$$E(y^{(g)} \mid \boldsymbol{x}^*) = \zeta_{\boldsymbol{x}^*}^{(g)} = \alpha_{01}A + \alpha_{02}A^2 + A\sum_{i=1}^{q} \alpha_i x_i + \sum_{i=1}^{q} \alpha_{ii}x_i^2 + \sum_{i<j=1}^{q} \alpha_{ij}x_i x_j, \tag{12.3.7}$$

where the experimental domain is given by

$$\Xi = \left\{ (A, x_1, x_2, \ldots, x_q) \mid A \in [A_L, A_U], A_L > 0, x_i \geq 0, i = 1(1)q, \sum_i x_i = 1 \right\}. \tag{12.3.8}$$

The assumption $A_L > 0$ ensures that some amount of the mixture is used in the experiment. A similar model has been suggested by Pal and Mandal (2012a) in the single-response case. The problem remains that of finding the optimum design for estimating the optimum mixing proportions and the optimum mixture amount for each of the responses.

Proceeding as in Sect. 12.3.2, the model (12.3.7) can be written as

$$\zeta_{\boldsymbol{x}^*}^{(g)} = \beta_{00}^{(g)}A^2 + A\sum_i \beta_{0i}^{(g)}x_i + \sum_i \beta_{ii}^{(g)}x_i^2 + \sum_{i<j} \beta_{ij}^{(g)}x_i x_j = \boldsymbol{f}'(\boldsymbol{x})\boldsymbol{\beta}^{(g)} = \boldsymbol{x}^{*\prime}B^{(g)}\boldsymbol{x}^*,$$

Table 12.10 Optimum designs for $q = 3, 4, \ldots, 8$; $p = 2, 3, 4$ and some combinations of $(\sigma^*_{gg}, v^{(g)}, w^{(g)})$s

q	p	$(\sigma^*_{11}, \ldots, \sigma^*_{pp})$	$(v^{(1)}, \ldots, v^{(p)})$	α	$\phi(\xi)$
3	2	(1, 3)	(0.2, 0.3)	0.3642	4452.166
	2	(1, 10)	(0.15, 0.3)	0.3656	12964.57
	3	(1, 3, 7)	(0.12, 0.20, 0.28)	0.3596	10251.99
	3	(3, 10, 15)	(0.2, 0.25, 0.3)	0.3637	30588.17
	4	(3, 10, 15, 8)	(0.2, 0.25, 0.3, 0.17)	0.3505	34643.97
	4	(3, 5, 10, 14)	(0.12, 0.25, 0.29, 0.33)	0.3653	37521.09
4	2	(1, 3)	(0.1, 0.23)	0.2835	20643.334
	2	(1, 10)	(0.07, 0.18)	0.2823	61395.741
	3	(1, 3, 7)	(0.08, 0.12, 0.20)	0.2714	44210.803
	3	(3, 10, 15)	(0.18, 0.23, 0.11)	0.2656	110973.88
	4	(3, 10, 15, 8)	(0.2, 0.21, 0.10, 0.07)	0.2650	109763.18
	4	(3, 5, 10, 14)	(0.12, 0.2, 0.15, 0.23)	0.2822	157399.69
5	2	(1, 3)	(0.1, 0.18)	0.2415	87805.970
	2	(1, 10)	(0.07, 0.18)	0.2389	204318.79
	3	(1, 3, 7)	(0.08, 0.12, 0.19)	0.2378	192034.41
	3	(3, 10, 15)	(0.18, 0.09, 0.11)	0.2277	314729.42
	4	(3, 10, 15, 8)	(0.1, 0.16, 0.14, 0.06)	0.2315	467097.13
	4	(3, 5, 10, 14)	(0.18, 0.13, 0.11, 0.07)	0.2245	333748.61
7	2	(1, 3)	(0.1, 0.12)	0.2391	50339.700
	2	(1, 10)	(0.04, 0.14)	0.2407	159253.90
	3	(1, 3, 7)	(0.08, 0.12, 0.13)	0.2400	148547.10
	3	(3, 10, 15)	(0.14, 0.09, 0.11)	0.2380	322630.00
	4	(3, 10, 15, 8)	(0.1, 0.16, 0.14, 0.06)	0.2401	493408.70
	4	(3, 5, 10, 14)	(0.1, 0.13, 0.11, 0.07)	0.2362	325363.60
8	2	(1, 3)	(0.1, 0.12)	0.2409	50931.790
	2	(1, 10)	(0.03, 0.11)	0.2406	135937.20
	3	(1, 3, 7)	(0.08, 0.12, 0.01)	0.2393	119435.00
	3	(3, 10, 15)	(0.11, 0.09, 0.07)	0.2368	245709.60
	4	(3, 10, 15, 8)	(0.1, 0.12, 0.03, 0.06)	0.2337	256769.40
	4	(3, 5, 10, 14)	(0.1, 0.05, 0.11, 0.07)	0.2369	283674.80

where $\boldsymbol{f}(\boldsymbol{x}^*) = (A^2, Ax_1, Ax_2, \ldots, Ax_q, x_1^2, x_2^2, \ldots, x_q^2, x_1x_2, x_1x_3, \ldots, x_{q-1}x_q)'$, $\boldsymbol{\beta}^{(g)} = (\beta^{(g)}_{00}, \beta^{(g)}_{01}, \ldots, \beta^{(g)}_{0q}, \beta^{(g)}_{11}\beta^{(g)}_{22}, \ldots, \beta^{(g)}_{qq}, \beta^{(g)}_{12}, \ldots, \beta^{(g)}_{q-1,q})'$, $\boldsymbol{B}^{(g)}$ is a symmetric matrix with $(i+1, i+1)$-th element $\beta^{(g)}_{ii}$, $i = 0, 1, \ldots, q$, and the $(i+1, j+1)$-th element $\beta^{(g)}_{ij}$, $i < j$, and the experimental region is given by (12.3.3). Under the assumption of concavity of $\zeta^{(g)}_{\boldsymbol{x}^*}$ and that the maximizing point $\zeta^{(g)}_{\boldsymbol{x}^*}$ is an interior point of (12.3.3), the optimum point for each $g = 1, 2, \ldots, p$ is $\boldsymbol{\gamma}^{(g)*} = \delta^{(g)-1}\boldsymbol{B}^{(g)-1}\boldsymbol{1}$, where $\delta^{(g)} = \boldsymbol{c}'\boldsymbol{B}^{(g)-1}\boldsymbol{c}$ and $\boldsymbol{c} = (0, 1, 1, \ldots, 1)'$. For any continuous design ξ, the dispersion matrix Disp$(\hat{\boldsymbol{\gamma}}^{(g)*})$ of the estimate $\hat{\boldsymbol{\gamma}}^{(g)*}$ is singular and is dependent on the model parameters so that, as in the earlier subsections, the pseudo-Bayesian approach of Pal and Mandal (2006) can be used todefine a parameter-free criterion function as $\phi(\xi) = \mathcal{E}[\text{Trace}\{\text{Disp}(\hat{\boldsymbol{\gamma}}^{(g)*})\}]$, where $\mathcal{E}$ denotes the expectation with respect to the prior distribution of $\boldsymbol{\gamma}^{(g)*}$, for which, arguing as in Pal and Mandal (2006), the second-order moments (pure and mixed) are defined as

$$\mathcal{E}(\gamma_1^{(g)*2}) = v^{(g)}, i = 1(1)p; \mathcal{E}(\gamma_i^{(g)*}\gamma_j^{(g)*}) = w^{(g)}, i \neq j = 1, 2, \ldots, q; v^{(g)} \geq w^{(g)} > 0;$$
$$\mathcal{E}(A^{(g)*2}) = a^{(g)}, \mathcal{E}(A^{(g)*}\gamma_i^{(g)*}) = b^{(g)}, i = 1, 2, \ldots, q, g = 1, 2, \ldots, p.$$

Clearly, since $\boldsymbol{c}'\boldsymbol{x}^* = 1$ and $B^{(g)}$ is negative definite, the prior moments satisfy $w^{(g)} \leq v^{(g)}, 1/q^2 < v^{(g)} < 1/q, qv^{(g)} + q(q-1)w^{(g)} = 1, g = 1, 2, \ldots, p$.

Mimicking the argument as in Sect. 12.3.2, one can initially confine the search for trace-optimal design within the subclass $\mathcal{D}_q$ of designs having support points and masses as given in Table 12.8, with $a_0 = 0$, and $w_{-1} = w_1$, $p_i = p_i''$, $i = 1, 2$. For $2 \leq q \leq 4$, Pal and Mandal (2013) and Mandal examined the optimality or otherwise of the trace-optimal designs in $\mathcal{D}_q$ within the whole class of competing designs using Theorem 12.3.1 (equivalence theorem), and showed them to be optimal. As the algebraic derivations were rather involved, the condition (12.3.6) was checked by numerical computation, using several points in the experimental region and for various combinations of $(\sigma_{gg}^*, g = 1, 2, \ldots, p)$ and the apriori moments $((a^{(g)}, v^{(g)}, w^{(g)}), g = 1, 2, \ldots, p)$. The designs for $q \geq 5$ can be similarly obtained and their optimality or otherwise checked numerically through Theorem 12.3.1. The following table is an excerpt from Pal and Mandal (2013) (Table 12.11).

12.4 Mixture Designs in Blocks

12.4.1 Preliminaries

Blocking is a desirable property of any response surface design as it controls the heterogeneity of the experimental units. It is also an important tool for obtaining increased precision of estimates of the parameters/parametric functions of interest. Suppose a mixture experiment on fertilizers is to be carried out at different locations. Groups of mixture blends may be used in different locations which may be called blocks. Some examples of mixture experiments where blocking is necessary are discussed in Cornell (2002). Mixture design with blocks was first considered by Nigam (1970) where the basic design was a symmetric simplex design as defined by Murty and Das (1968), and the mixture model was Scheffé's second-degree model. Blocked mixture designs may be categorized into three groups, viz. (i) orthogonal block designs based on symmetric simplex mixture designs, (ii) orthogonal block designs using mates of latin squares, and (iii) D-optimal minimum support designs in blocks (Goos and Donev 2007). In each case, Scheffé's quadratic mixture model is considered.

Table 12.11 Optimum designs for $q = 2, 3, 4$; $p = 2, 3, 4$ and some combinations of $(\sigma^*_{gg}, a^{(g)}, v^{(g)}, w^{(g)})$s

q	p	$(a^{(1)}, \ldots, a^{(p)})$	$(\sigma^*_{11}, \ldots, \sigma^*_{pp})$	$(v^{(1)}, \ldots, v^{(p)})$	p_1	p_1'	w_1	w_0	Trace
2	2	(0.2, 0.4)	(1, 3)	(0.30, 0.40)	0.29915	0.27692	0.31869	0.36261	122.780
	2	(0.2, 0.6)	(1, 10)	(0.27, 0.45)	0.29609	0.27151	0.31246	0.37507	482.261
	3	(0.2, 0.6, 0.8)	(1, 3, 7)	(0.27, 0.32, 0.40)	0.31207	0.28381	0.30335	0.39329	407.639
	3	(0.1, 0.5, 0.7)	(1, 6, 10)	(0.30, 0.37, 0.45)	0.29988	0.27424	0.37908	0.38043	706.915
	4	(0.1, 0.3, 0.5, 0.7)	(1, 4, 8, 10)	(0.28, 0.32, 0.4, 0.48)	0.29785	0.27321	0.31226	0.37548	934.674
	4	(0.1, 0.5, 0.7, 0.9)	(1, 6, 10, 14)	(0.3, 0.37, 0.45, 0.26)	0.32015	0.29093	0.30062	0.39876	1075.68
3	2	(0.2, 0.4)	(1, 3)	(0.15, 0.25)	0.12969	0.10904	0.35186	0.29628	973.012
	2	(0.2, 0.6)	(1, 10)	(0.24, 0.32)	0.13363	0.11377	0.35000	0.30000	4301.49
	3	(0.2, 0.6, 0.8)	(1, 3, 7)	(0.14, 0.24, 0.32)	0.13552	0.11149	0.34406	0.31188	3902.46
	3	(0.1, 0.5, 0.7)	(1, 6, 10)	(0.12, 0.20, 0.33)	0.13423	0.11132	0.34596	0.30807	5647.98
	4	(0.1, 0.3, 0.5, 0.7)	(1, 4, 8, 10)	(0.12, 0.18, 0.22, 0.30)	0.13339	0.10947	0.34554	0.30892	6562.84
	4	(0.1, 0.5, 0.7, 0.9)	(1, 6, 10, 14)	(0.15, 0.20, 0.25, 0.32)	0.13624	0.11049	0.34218	0.31564	10489.7
4	2	(0.6, 0.2)	(1, 3)	(0.08, 0.0, 0.18)	0.07387	0.05906	0.37747	0.24505	4118.26
	2	(0.2, 0.4)	(1, 10)	(0.10, 0.15)	0.07456	0.05640	0.37262	0.25476	10645.4
	3	(0.2, 0.6, 0.8)	(1, 3, 7)	(0.07, 0.14, 0.24)	0.07877	0.05887	0.36902	0.26196	16029.4
	3	(0.1, 0.5, 0.7)	(1, 6, 10)	(0.08, 0.18, 0.22)	0.07880	0.05888	0.36899	0.26202	24836.1
	4	(0.1, 0.3, 0.5, 0.7)	(1, 4, 8, 10)	(0.07, 0.12, 0.18, 0.22)	0.07724	0.05899	0.37132	0.25735	29931.0
	4	(0.1, 0.5, 0.7, 0.9)	(1, 6, 10, 14)	(0.09, 0.15, 0.20, 0.24)	0.07913	0.05914	0.36879	0.26242	47361.1

Let N mixture blends be allocated to t blocks, where the w-th block contains n_w blends, $w = 1, 2, \ldots, t$; $n_1 + n_2 + \cdots + n_t = N$. It is to be noted that the blends in a block may not be all distinct.

Scheffé's quadratic mixture model under blocking may be written as

$$y_u = \sum_{i=1}^{q} \beta_i x_{iu} + \sum_{1 \le i < j \le q}^{q} \beta_{ij} x_{iu} x_{ju} + \sum_{w=1}^{t} \gamma_w (z_{wu} - \bar{z}_w) + e_u, \qquad (12.4.1)$$

where x_{iu}s are the proportion of the ith mixture component for the uth blend; $x_{iu} \ge 0$, $\sum_i x_{iu} = 1 \,\forall\, u$; γ_w is the effect of the wth block, z_{wu} is an indicator variable assuming the value 1 or 0 according as the uth blend is in the wth block or not; $\bar{z} = \frac{1}{N} \Sigma_u z_{uw}$ and e_u is the error associated with y_u, the observation from the uth blend, $u = 1, 2, \ldots, N$.

In matrix notation, the model (12.4.1) can be written as

$$\boldsymbol{y} = X\boldsymbol{\beta} + (Z - \bar{Z})\boldsymbol{\gamma} + \boldsymbol{e}, \qquad (12.4.2)$$

where X and $(Z - \bar{Z})$ are the coefficient matrices of βs and γs, respectively, with $\bar{Z} = \left(\frac{n_1}{N}\mathbf{1}_N, \frac{n_2}{N}\mathbf{1}_N, \ldots, \frac{n_t}{N}\mathbf{1}_N\right)$, $\mathbf{1}_N = (1, 1, \ldots, 1)'$.

It is desirable to estimate the β-parameters independently of the block effects. A design where this feature is realized is called a mixture design with orthogonal blocks. In this connection, it may be mentioned that a linear function of γs is estimable provided it is a contrast in γs. We restrict to the class $\mathcal{D}_B$ of mixture designs where all βs and all block contrasts are estimable.

For the model (12.4.2), it can shown that a necessary and sufficient condition for orthogonal blocking is

$$X'(Z - \bar{Z}) = 0 \qquad (12.4.3)$$

which is equivalent to

$$\sum_{u \in B_w} x_{iu} = k_i n_w, \quad \sum_{u \in B_w} x_{iu} x_{ju} = k_{ij} n_w \qquad (12.4.4)$$

where B_w denotes the wth block, k_i and k_{ij} denote some constants, $w = 1, 2, \ldots, t$ and $1 \le i < j \le q$.

Nigam (1970, 1976) considered blocking of the symmetric simplex designs (Murty and Das 1968) for mixture experiments, imposing some conditions on the mixtures. Later, Singh et al. (1982) revised the conditions of Nigam to have the blocks orthogonal by taking constants k_i and k_{ij} in (12.4.4) independent of the factors. John (1984) considered mixture designs with orthogonal blocks modifying the conditions in (12.4.4) for equal block size. Draper et al. (1993), Lewis et al. (1994), and others considered mixture designs with orthogonal blocks ensuring John's conditions. Also, Prescott (2000) constructed such designs ensuring the condition (12.4.4) for both equal and unequal block sizes using projection designs. Actually, it is seen that the conditions of both Singh et al. (1982) and John (1984) are sufficient and can be obtained as particular cases of the condition (12.4.4).

Deviating from orthogonal block designs, Goos and Donev (2007) studied another kind of blocked mixture designs which are D-optimal in the class of minimum support designs.

12.4.2 *Orthogonal Mixture Designs Based on Symmetric Simplex Block Designs*

As mentioned earlier, Nigam (1970) proposed block designs for mixture experiments, considering the symmetric simplex design (Murty and Das 1968) for which the moments of different orders are invariant with respect to the components.

To get the estimates of the regression parameters β_is and β_{ij}s from the normal equations, adjusting for the block effects, Nigam (1970, 1976) imposed the following blocking conditions

$$\sum_{u=1}^{n_w} x_{iu} = k_{1w},\ \sum_{u=1}^{n_w} x_{iu}x_{ju} = k_{2w},\ i \neq j = 1, 2, \ldots, q \qquad (12.4.5)$$

However, a design satisfying (12.4.5) may not necessarily be orthogonal.

For orthogonal estimation of the mixture parameters, Singh et al. (1982) revised the condition (12.4.5) as

$$k_{1w} = k_1 n_w,\ k_{2w} = k_2 n_w, \quad w = 1, 2, \ldots, t. \qquad (12.4.6)$$

It is noted that the condition (12.4.6) is a particular case of (12.4.4) when $k_i = k_1$ and $k_{ij} = k_2$, $1 \le i < j \le q$.

The following design (Cornell 2002) is a symmetric simplex design with orthogonal blocks satisfying the conditions (12.4.6) with $n_1 = n_2 = 6$, $k_1 = 2$, $k_2 = \frac{1}{4}$ when $q = 3$.

Example 12.4.1

Block 1			Block II		
1	0	0	1	0	0
0	1	0	0	1	0
0	0	1	0	0	1
$\frac{1}{2}$	$\frac{1}{2}$	0	$\frac{4}{6}$	$\frac{1}{6}$	$\frac{1}{6}$
$\frac{1}{2}$	0	$\frac{1}{2}$	$\frac{1}{6}$	$\frac{4}{6}$	$\frac{1}{6}$
0	$\frac{1}{2}$	$\frac{1}{2}$	$\frac{1}{6}$	$\frac{1}{6}$	$\frac{4}{6}$

12.4.3 Mixture Designs with Orthogonal Blocks Based on Mates of Latin Squares

John (1984) made use of the following conditions

$$\sum_{u \in \zeta_w} x_{iu} = k_i, \quad \sum_{u \in \zeta_w} x_{iu} x_{ju} = k_{ij}, \quad 1 \le i \ne j \le q \tag{12.4.7}$$

for orthogonal blocking.

It follows from (12.4.4) that (12.4.7) results as particular cases of (12.4.4) when the block sizes are equal.

John (1984), Draper et al. (1993), Lewis et al. (1994), and others constructed mixture designs with orthogonal blocks using mates of Latin squares. Prescott (2000) used projection designs for mixture experiments to construct designs with orthogonal blocks ensuring (12.4.4).

Two Latin squares of side q are said to be *mates* of each other if the sum of products of the elements of ith and jth columns from each square is the same, $1 \le i < j \le q$.

For a discussion on mates of 4×4 Latin squares, one can see Draper et al. (1993) and also Cornell (2002). Consider the following two 4×4 Latin squares

$$L_1 = \begin{pmatrix} a & b & c & d \\ b & c & d & a \\ c & d & a & b \\ d & a & b & c \end{pmatrix}, \quad L_2 = \begin{pmatrix} a & d & c & b \\ b & a & d & c \\ c & b & a & d \\ d & c & b & a \end{pmatrix}$$

It is easy to see that the sum of cross-products of the elements of columns 1 and 2 from each of L_1 and L_2 is the same. Similarly, it can be verified that this is true for any pair of columns of L_1 and L_2. So, it follows that L_1 and L_2 are mates of each other.

By assuming the rows of a square as blends, it is easily seen that the mates of Latin squares can effectively be used to construct mixture designs with orthogonal blocks satisfying John's conditions. Also, as the blends in the blocks are the rows of Latin squares (side q), it follows that

$$\sum_{1 \le i < j \le q} x_{iu} x_{ju} = \rho \sum_{i=1}^{q} x_{iu}, \quad 1 \le u \le N \tag{12.4.8}$$

where ρ is a constant. So, the coefficient matrix X for the mixture parameters is not of full column rank. As a remedy, at least one mixture point may be included in each block so that singularity is removed and, at the same time, the conditions in (12.4.7) are satisfied. Usually, the centroid point having equal proportion for each component is used.

Mixture designs with two orthogonal blocks were constructed in Draper et al. (1993) using mates of 4×4 Latin squares. The proportions indicated by the 4 symbols a, b, c and d are arbitrary satisfying $a, b, c, d \ge 0$ and $a + b + c + d = 1$.

Draper et al. (1993) computed their values by maximizing the determinant of $(X'X)$. Lewis et al. (1994) developed some general methods of constructing mates of Latin squares and proposed methods of constructing designs for three or more factors in two or more blocks.

Draper et al. (1993) constructed the following design with two orthogonal blocks each of size 9 using two pairs of mates of 4 × 4 Latin squares.

Example 12.4.2

$$\begin{array}{cc} \text{Block I} & \text{Block II} \\ \begin{pmatrix} a & b & c & d \\ b & a & d & c \\ c & d & b & a \\ d & c & a & b \\ \frac{1}{4} & \frac{1}{4} & \frac{1}{4} & \frac{1}{4} \\ a & b & c & d \\ b & c & d & a \\ c & d & a & b \\ d & a & b & c \end{pmatrix} & \begin{pmatrix} a & b & d & c \\ b & a & c & d \\ c & d & a & b \\ d & c & b & a \\ \frac{1}{4} & \frac{1}{4} & \frac{1}{4} & \frac{1}{4} \\ a & d & c & b \\ b & a & d & c \\ c & b & a & d \\ d & c & b & a \end{pmatrix} \end{array}$$

The values for the proportions were computed as $a = 0.24, b = 0, c = 0, d = 0.76$.

12.4.4 D-Optimal Minimum Support Mixture Designs

Minimum support mixture design is a design which uses only that many mixture combinations as there are parameters in the model. In Scheffé's quadratic model, there are $m = q(q+1)/2$ parameters. So, a minimum support design uses only m distinct support points where q_i observations are taken from the ith support point, $i = 1, 2, \ldots m$; $\Sigma_{i=1}^{m} q_i = N$. Goos and Donev (2007) considered D-optimal design in the class of minimum support designs. They also compared these designs with comparable orthogonal designs and with the designs which were D-optimal for the whole class, not merely optimal in the class of minimum support designs.

12.4.5 Model and Estimators

Though Goos and Donev (2007) assumed a general setup, we describe the procedure with respect to the present mixture model setup.

Let us consider the model (12.4.2) with a little modification by dropping $\bar{Z}$, as

$$\boldsymbol{y} = \boldsymbol{X\beta} + \boldsymbol{Z\gamma} + \boldsymbol{e}. \qquad (12.4.9)$$

It follows that the design matrix $(\boldsymbol{X}, \boldsymbol{Z})$ is not of full column rank as the sum of the columns corresponding to the linear regression parameters is equal to that of the columns in $\boldsymbol{Z}$. To make the coefficient matrix non-singular, Goos and Donev (2007) modified the model by dropping the last column of $\boldsymbol{Z}$ and the corresponding parameter γ_t.

Define

$$F = (\boldsymbol{X}, \boldsymbol{Z}_0), \quad \boldsymbol{\theta}' = (\boldsymbol{\beta}', \boldsymbol{\gamma}'_0) \tag{12.4.10}$$

where

$$\boldsymbol{Z}_0 = (Z_1, Z_2, \ldots, Z_{t-1}), \quad \boldsymbol{\gamma}'_0 = (\gamma_1, \gamma_2, \ldots, \gamma_{t-1}), \tag{12.4.11}$$

Z_i being the ith column of $\boldsymbol{Z}_0$, $i = 1, 2, \ldots, t-1$.

So, the modified model can be written as

$$\boldsymbol{y} = \boldsymbol{F\theta} + \boldsymbol{e} \tag{12.4.12}$$

and the least squares estimator of $\boldsymbol{\theta}$ is given by

$$\hat{\boldsymbol{\theta}} = (\boldsymbol{F}'\boldsymbol{F})^{-1}\boldsymbol{F}'\boldsymbol{y} \tag{12.4.13}$$

with

$$\text{Disp}(\hat{\boldsymbol{\theta}}) = \sigma^2(\boldsymbol{F}'\boldsymbol{F})^{-1} \tag{12.4.14}$$

12.4.6 Construction of Optimum Designs

Goos and Donev (2007) proposed a two-stage method for the construction of D-optimal minimum support design for the estimation of the mixture parameters in the presence of the block effects. They made use of the following theorem due to Donev (1989):

Theorem 12.4.1

$$\mid \boldsymbol{F}'\boldsymbol{F} \mid = \mid \boldsymbol{X}'\boldsymbol{X} \mid . \mid \boldsymbol{A} \mid \tag{12.4.15}$$

where

$$\mid \boldsymbol{A} \mid = \mid (a_{ij}) \mid^{(t-1)\times(t-1)}, \quad a_{ii} = n'_i - \sum_{i \in D_i} q_i^{-1}, \quad a_{ij} = -\sum_{i \in D_{ij}} q_i^{-1} \tag{12.4.16}$$

n'_i = *number of points in the ith block that appear more than once in the entire design,*
D_i = *set of the design points in the ith block which occurs in other blocks also and* $D_{ij} = D_i \cap D_j$; $i \neq j = 1, 2, \ldots, t$.

$\mid \cdot \mid$ *denotes the determinant of the matrix* $(\cdot)$.

Note that the elements of the matrix $\boldsymbol{A}$ depend only on the assignment of the design points to the blocks. The expression of $\boldsymbol{A}$ is valid only if no design point is replicated within a block; its replication, if any, should occur in other blocks.

From (12.4.15), it follows that $\mid \boldsymbol{F}'\boldsymbol{F} \mid$ should be maximized for getting the *D*-optimal minimum support design. For this, we have to proceed through the following two stages.

(i) Choose $m = q(q+1)/2$ distinct support points and replicate them as evenly as possible to get N runs so that $\mid \boldsymbol{X}'\boldsymbol{X} \mid$ is maximized. For this, each of the m support points should occur $\left[\frac{n}{m}\right]$ or $\left[\frac{n}{m}\right]+1$ times in the design. It is immaterial which one is replicated most.
(ii) Looking at $\mid \boldsymbol{A} \mid$ in (12.4.16), it follows that the chosen support points should be distributed as evenly as possible over the blocks avoiding more than one occurrence of any point in any block. The assignment of the non-replicated design points to the blocks does not affect $\mid \boldsymbol{A} \mid$. So, the design may not be unique in respect of *D*-optimality criterion.

Example 12.4.3 We present an example of *D*-optimal minimum support design for quadratic mixture model with 4 factors as illustrated by Goos and Donev (2007).

For estimating the 10 parameters in the Scheffé's quadratic model with 4-factors, the ten support points of the (4, 2) simplex design are chosen. For a design with 18 runs in 2 blocks each containing 9 runs, 8 of the 10 support points should be used twice and the remaining two just once. This will maximize $\mid \boldsymbol{X}'\boldsymbol{X} \mid$. The 8 replicated support points should occur in both the blocks and the remaining two should occur in different blocks. It does not matter which eight support points are duplicated. One such *D*-optimal design is the following:

Block 1	Block II
$0\ 1\ 0\ 0$	$1\ 0\ 0\ 0$
$0\ 0\ 1\ 0$	$0\ 0\ 1\ 0$
$0\ 0\ 0\ 1$	$0\ 0\ 0\ 1$
$\frac{1}{2}\ \frac{1}{2}\ 0\ 0$	$\frac{1}{2}\ \frac{1}{2}\ 0\ 0$
$\frac{1}{2}\ 0\ 0\ \frac{1}{2}$	$\frac{1}{2}\ 0\ 0\ \frac{1}{2}$
$0\ 0\ \frac{1}{2}\ \frac{1}{2}$	$0\ 0\ \frac{1}{2}\ \frac{1}{2}$
$0\ 0\ \frac{1}{2}\ \frac{1}{2}$	$0\ \frac{1}{2}\ \frac{1}{2}\ 0$
$0\ \frac{1}{2}\ \frac{1}{2}\ 0$	$0\ \frac{1}{2}\ 0\ \frac{1}{2}$
$\frac{1}{2}\ 0\ \frac{1}{2}\ 0$	$\frac{1}{2}\ 0\ \frac{1}{2}\ 0$

Goos and Donev (2007) also discussed the cases when the model involved (i) one blocking variable with random effects, (ii) two blocking variables with fixed effects each, and (iii) two blocking variables with random effects each.

12.4.7 Comparison with Other Designs

Goos and Donev (2007). compared the following two mixture designs D_1 and D_2 each with three factors having 8 runs divided into two equal blocks

$$
\begin{array}{cc|cc}
\multicolumn{2}{c}{D_1} & \multicolumn{2}{c}{D_2} \\
\text{Block 1} & \text{Block 2} & \text{Block 1} & \text{Block 2} \\
1\ 0\ 0 & 0\ 1\ 0 & \frac{1}{2}\ \frac{1}{2}\ 0 & \frac{1}{6}\ \frac{2}{3}\ \frac{1}{6} \\
0\ 0\ 1 & \frac{1}{2}\ \frac{1}{2}\ 0 & \frac{1}{2}\ 0\ \frac{1}{2} & \frac{2}{3}\ \frac{1}{6}\ \frac{1}{6} \\
\frac{1}{2}\ \frac{1}{2}\ 0 & 0\ \frac{1}{2}\ \frac{1}{2} & 0\ \frac{1}{2}\ \frac{1}{2} & \frac{1}{6}\ \frac{1}{6}\ \frac{2}{3} \\
0\ \frac{1}{2}\ \frac{1}{2} & \frac{1}{2}\ 0\ \frac{1}{2} & \frac{1}{3}\ \frac{1}{3}\ \frac{1}{3} & \frac{1}{3}\ \frac{1}{3}\ \frac{1}{3}
\end{array}
$$

where D_1 is a D-optimal minimum support design and D_2 is a mixture design with orthogonal blocks (Prescott 2000). The relative D-efficiency of D_1 with respect to D_2 computed through $\{\mid \boldsymbol{F}_1'\boldsymbol{F}_1 \mid / \mid \boldsymbol{F}_2'\boldsymbol{F}_2 \mid\}^{1/6}$ exceeds 3, where $\boldsymbol{F}_1$ and $\boldsymbol{F}_2$ are the model matrices for D_1 and D_2, respectively, as described in (12.4.12).

Consider the following design D_3 in two blocks. This is D-optimal in the entire class obtained through grid search.

Block 1	Block 2
(0.6 0.4 0)	0.4 0.6 0
(0.5 0 0.5)	0 0.6 0.4
(0 1 0)	1 0 0
(0 0.4 0.6)	0 0 1

It is observed that the relative D-efficiency of D_1 with respect to D_3 is less than 3.11 %.

Remark 12.4.1 The design D_2 proposed by Prescott (2000) can be seen to belong to the following class of designs with two blocks:

$$\begin{array}{ccc} & \boldsymbol{B}_1 & \\ \alpha & \frac{1-\alpha}{2} & \frac{1-\alpha}{2} \\ \frac{1-\alpha}{2} & \alpha & \frac{1-\alpha}{2} \\ \frac{1-\alpha}{2} & \frac{1-\alpha}{2} & \alpha \\ \frac{1}{3} & \frac{1}{3} & \frac{1}{3} \end{array} \qquad \begin{array}{ccc} & \boldsymbol{B}_2 & \\ \beta & \frac{1-\beta}{2} & \frac{1-\beta}{2} \\ \frac{1-\beta}{2} & \beta & \frac{1-\beta}{2} \\ \frac{1-\beta}{2} & \frac{1-\beta}{2} & \beta \\ \frac{1}{3} & \frac{1}{3} & \frac{1}{3} \end{array}$$

where

$$\beta = \frac{2}{3} - \alpha, \alpha \in [0, \frac{1}{3}).$$

It is readily seen that D_2 is obtained when $\alpha = 0$.

Remark 12.4.2 Prescott (2000) also considered a mixture design with 3 factors in 3 orthogonal blocks $\boldsymbol{B}_1$, $\boldsymbol{B}_2$, and $\boldsymbol{B}_3$ of sizes 6, 6, and 8, respectively, through projections. It can be seen that Prescott's design belongs to the following class of designs with three blocks:

$$\begin{array}{ccc} & \boldsymbol{B}_1 & \\ \alpha & \frac{1-\alpha}{2} & \frac{1-\alpha}{2} \\ \frac{1-\alpha}{2} & \alpha & \frac{1-\alpha}{2} \\ \frac{1-\alpha}{2} & \frac{1-\alpha}{2} & \alpha \\ \frac{1}{3} & \frac{1}{3} & \frac{1}{3} \\ \frac{1}{3} & \frac{1}{3} & \frac{1}{3} \\ \frac{1}{3} & \frac{1}{3} & \frac{1}{3} \end{array} \qquad \begin{array}{ccc} & \boldsymbol{B}_2 & \\ \beta & \frac{1-\beta}{2} & \frac{1-\beta}{2} \\ \frac{1-\beta}{2} & \beta & \frac{1-\beta}{2} \\ \frac{1-\beta}{2} & \frac{1-\beta}{2} & \beta \\ \frac{1}{3} & \frac{1}{3} & \frac{1}{3} \\ \frac{1}{3} & \frac{1}{3} & \frac{1}{3} \\ \frac{1}{3} & \frac{1}{3} & \frac{1}{3} \end{array} \qquad \begin{array}{ccc} & \boldsymbol{B}_3 & \\ \delta & \frac{1-\delta}{2} & \frac{1-\delta}{2} \\ \frac{1-\delta}{2} & \delta & \frac{1-\delta}{2} \\ \frac{1-\delta}{2} & \frac{1-\delta}{2} & \delta \\ \gamma & \frac{1-\gamma}{2} & \frac{1-\gamma}{2} \\ \frac{1-\gamma}{2} & \gamma & \frac{1-\gamma}{2} \\ \frac{1-\gamma}{2} & \frac{1-\gamma}{2} & \gamma \\ \frac{1}{3} & \frac{1}{3} & \frac{1}{3} \\ \frac{1}{3} & \frac{1}{3} & \frac{1}{3} \end{array}$$

where

$$\beta = \frac{2}{3} - \alpha, \delta = \frac{1}{3} + x, \gamma = \frac{1}{3} - x, \alpha \in [0, \frac{1}{3}), x = \left(\alpha - \frac{1}{3}\right)\frac{\sqrt{2}}{\sqrt{3}}$$

The design considered in Prescott (2000) is obtained when $\alpha = 0$.

Several such mixture designs in orthogonal blocks, as generalizations of those in Prescott (2000), are currently under investigation. Das and Sinha (2014).

References

Atkinson, A. C., & Fedorov, V. V. (1975a). The design of experiments for discriminating between two rival models. *Biometrika, 62*, 57–70.

Atkinson, A. C., & Fedorov, V. V. (1975b). Optimal design: Experiments for discriminating between several models. *Biometrika, 62*, 289–303.

Box, G. E. P., & Draper, N. R. (1959). A basis for the selection of response surface design. *Journal of the American Statistical Association, 54*, 622–654.

Chakrabarti, H., & Mandal, N. K. (1995). Mixture experiments: All-variance and all-bias designs. *Bulletin of the Calcutta Statistical Association, 45*, 219–234.

Cornell, J. A. (2002). *Experiments with mixtures* (3rd ed.). New York: Wiley.

Darroch, J. N., & Waller, J. (1985). Additivity and interaction in three-component experiments with mixtures. *Biometrika, 72*, 153–163.

Das, P., & Sinha, B. K. (2014). Mixture designs in orthogonal blocks. Under preparation.

Dette, H. (1994). Discrimination designs for polynomial regression on compact intervals. *The Annals of Statistics, 22*, 890–903.

Dette, H., & Titoff, S. (2009). Optimal discrimination designs. *The Annals of Statistics, 37*, 2056–2082.

Dette, H., Melas, V. B., & Shpilev, P. (2012). T-optimal designs for discrimination between two polynomial models. *The Annals of Statistics, 40*, 188–205.

Donev, A. N. (1989). Design of experiments with both mixture and qualitative factors. Journal of the Royal Statistical Society. *Series B (Methodological), 51*, 297–302.

Draper, N. R., & Lawrence, W. E. (1965a). Mixture designs for three factors. Journal of the Royal Statistical Society. *Series B (Methodological), 27*, 450–465.

Draper, N. R., & Lawrence, W. E. (1965b). Mixture designs for four factors. Journal of the Royal Statistical Society. *Series B (Methodological), 27*, 473–478.

Draper, N. R., Prescott, P., Lewis, S. M., Dean, A. M., John, P. W. M., & Tuck, M. G. (1993). Mixture designs for four components in orthogonal blocks. *Technometrics, 35*, 268–276.

Draper, N. R., & Pukelsheim, F. (1999). Kiefer ordering of simplex designs for first and second degree mixture models. *Journal of Statistical Planning and Inference, 79*, 325–348.

Draper, N. R., Heiligers, B., & Pukelsheim, F. (2000). Kiefer ordering of simplex designs for second-degree mixture models with four or more ingredients. *Annals of Statistics, 28*, 578–590.

Fedorov, V., & Khabarov, V. (1986). Duality of optimal designs for model discrimination and parameter estimation. *Biometrika, 73*, 183–190.

Galil, Z., & Kiefer, J. (1977). Comparison of simplex designs for quadratic mixture models. *Technometrics, 19*, 445–453.

Goos, P., & Donev, A. N. (2007). D-optimal minimum support mixture designs in blocks. *Metrika, 65*, 53–68.

John, P. W. M. (1984). *Experiments with Mixtures involving Process Variables, Technical Report 8*. Austin, Center for Statistical Sciences : The University of Texas.

Lewis, S. M., Dean, A. M., Draper, N. R., & Prescott, P. (1994). Mixture designs for q components in orthogonal blocks. Journal of the Royal Statistical Society. *Series B (Methodological), 56*, 457–467.

Liski, E. P., Luoma, A., Mandal, N. K., & Sinha, B. K. (1997). Optimal design for an inverse prediction problem under random coefficient regression models. Journal of the Indian Society of Agricultural Statistics, XLIX, 277–288.

Liski, E. P., Mandal, N. K., Shah, K. R., & Sinha, B. K. (2002). *Topics in optimal design* (Vol. 163). Lecture notes in statistics New York: Springer.

Mandal, N. K., & Pal, M. (2012). Optimum designs for estimation of optimum amount and proportions in a quadratic mixture-amount model. *Communications in Statistics-Theory and Methods, 41*, 1989–1999.

Mandal, N. K., & Pal, M. (2013). Optimum designs for optimum mixtures in multiresponse experiments. *Communications in Statistics-Simulation and Computation, 42*, 1104–1112.

Murty, J. A., & Das, M. N. (1968). Design and analysis of experiments with mixtures. *The Annals of Mathematical Statistics, 39*, 1517–1539.
Nigam, A. K. (1970). Block designs for mixture experiments. *The Annals of Mathematical Statistics, 41*, 1861–1869.
Nigam, A. K. (1976). Corrections to blocking conditions for mixture experiments. *The Annals of Statistics, 4*, 1294–1295.
Pal, M., & Mandal, N. K. (2006). Optimum designs for optimum mixtures. *Statistics & Probability Letters, 76*, 1369–1379.
Pal, M., & Mandal, N. K. (2007). Optimum mixture design via equivalence theorem. *Journal of Combinatorics, Information & System Sciences, 32*, 107–126.
Pal, M., Mandal, N. K., Sinha, B. K., & Das, P. (2010). Optimum mixture designs under random regression coefficients. *Journal of Applied Probability & Statistics, 5*, 15–29.
Pal, M., & Mandal, N. K. (2012a). Optimum designs for estimation of parameters in a quadratic mixture-amount model. *Communications in Statistics-Theory and Methods, 41*, 665–673.
Pal, M., & Mandal, N. K. (2012b). D-optimal designs for an additive quadratic mixture model with random regression coefficients. *Journal of the Indian Society of Agricultural Statistics, 66*, 441–446.
Pal, M., & Mandal, N. K. (2013). Optimum designs for optimum mixture and amount in multiresponse experiments. *The Thailand Statistician, 11*, 143–158.
Prescott, P. (2000). Projection designs for mixture experiments in orthogonal blocks. *Communications in Statistics-Theory and Methods, 29*, 2229–2253.
Rao, C. R. (1967). Least squares theory using an estimated dispersion matrix and its applications to measurement of signals. *Proceedings Fifth Berkeley Symposium on Mathematics and Probability, 1*, 355–372.
Roy, S. N., Gnanadesikan, R., & Srivastava, J. N. (1971). *Analysis and design of certain quantitative multiresponse experiments*. New York: Pergamon Press.
Singh, S. P., Pratap, M., & Das, M. N. (1982). Analysis of mixture experiments in presence of block effects. *Sankhya B, 44*, 270–277.
Tommasi, C., & López-Fidalgo, J. (2010). Bayesian optimum designs for discriminating between models with any distribution. *Computational Statistics & Data Analysis, 54*, 143–150.
Tsai, M. H., & Zen, M. M. (2004). Criterion robust optimal designs for model discrimination and parameter estimation: multivariate polynomial regression case. *Statistica Sinica, 14*, 591–601.
Wiens, D. P. (2009). Robust discrimination designs. *Journal of the Royal Statistical Society: Series B (Statistical Methodology), 71*, 805–829.

Bibliography

Atkinson, A. C. (1972). Planning experiments to detect inadequate regression models. *Biometrika, 5*, 275–294.

Atkinson, A. C., & Fedorov, V. V. (1975a). The design of experiments for discriminating between two rival models. *Biometrika, 62*, 57–70.

Atkinson, A. C., & Fedorov, V. V. (1975b). Optimal design: Experiments for discriminating between several models. *Biometrika, 62*, 289–303.

Biedermann, S., & Dette, H. (2001). Optimal designs for testing the functional form of a regression via nonparametric estimation techniques. *Statistics & Probability Letters, 52*, 215–224.

Box, G. E. P., & Cox, D. R. (1964). An analysis of transformations. *Journal of the Royal Statistical Society, Series B, 26*, 211–252.

Box, G. E. P., & Draper, N. R. (1959). A basis for the selection of response surface design. *Journal of the American Statistical Association, 54*, 622–654.

Chan, L. Y., & Sandhu, M. K. (1999). Optimal orthogonal block designs for a quadratic mixture model for three components. *Journal of Applied Statistics, 26*, 19–34.

Draper, N. R., & St. John, R. C. , (1977a). A mixture model with inverse terms. *Technometrics, 19*, 37–46.

Goos, P., & Donev, A. N. (2007). D-optimal minimum support mixture designs in blocks. *Metrika, 65*, 53–68.

Lambrakis, D. P. (1969). Experiments with mixtures: estimated regression function of the multiple-lattice design. Journal of the Royal Statistical Society. *Series B (Methodological), 31*, 276–284.

Minyu, Xie, & Zhaohai, Li. (2003). The optimal design for hypothesis test and its application in genetic linkage analysis. *Science in China Series A Mathematics, 46*, 737–749.

Pal, M., Mandal, N. K., Sinha, B. K., & Das, P. (2010). Optimum mixture designs under random regression coefficients. *Journal of Applied Probability & Statistics, 5*, 15–29.

Prescott, P., & Draper, N. R. (1998). Mixture designs for constrained components in orthogonal blocks. *Journal of Applied Statistics, 25*, 613–638.

Prescott, P., & Draper, N. R. (2004). Mixture component-amount designs via projections, including orthogonally blocked designs. *Journal of Quality Technology, 36*, 413–431.

Prescott, P., Draper, N. R., Lewis, S. M., & Dean, A. M. (1997). Further properties of mixture designs for five components in orthogonal blocks. *Journal of Applied Statistics, 24*, 147–156.

Saxena, S. K., & Nigam, A. K. (1973). Symmetric-simplex block designs for mixtures. Journal of the Royal Statistical Society. *Series B (Methodological), 35*, 466–472.

B. K. Sinha et al., *Optimal Mixture Experiments*, Lecture Notes in Statistics 1028,
DOI: 10.1007/978-81-322-1786-2,

Sinha, B. K., Das, P., Mandal, N. K., & Pal, M. (2010). Paramter estimation in linear and quadratic mixture models: A review. *Pakistan Journal of Statistics*, *26*, 77–96.

Tsai, M. H., & Zen, M. M. (2004). Criterion robust optimal designs for model discrimination and parameter estimation: multivariate polynomial regression case. *Statistica Sinica*, *14*, 591–601.

Zaigraev, A., & Liski, E. P. (2001). A stochastic characterization of Loewner optimality design criterion in linear models. *Metrika*, *53*, 207–222.

Author Index

B. K. Sinha et al., *Optimal Mixture Experiments*, Lecture Notes in Statistics 1028,
DOI: 10.1007/978-81-322-1786-2, © Springer India 2014

Subject Index

B. K. Sinha et al., *Optimal Mixture Experiments*, Lecture Notes in Statistics 1028,
DOI: 10.1007/978-81-322-1786-2, © Springer India 2014

MIX
Papier aus verantwortungsvollen Quellen
Paper from responsible sources
FSC® C105338
FSC
www.fsc.org

If you have any concerns about our products,
you can contact us on
ProductSafety@springernature.com

In case Publisher is established outside the EU,
the EU authorized representative is:
Springer Nature Customer Service Center GmbH
Europaplatz 3, 69115 Heidelberg, Germany

Printed by Libri Plureos GmbH
in Hamburg, Germany